DARWIN'S PICTURES

J U L I A V O S S

Translated by Lori Lantz

Darwin's Pictures

Views of Evolutionary Theory, 1837–1874

Yale UNIVERSITY PRESS

NEW HAVEN & LONDON

Geisteswissenschaften International—Translation Funding for Humanities and Social Sciences from Germany
A joint initiative of the Fritz Thyssen Foundation, the German Federal Foreign Office, and the German Publishers & Booksellers Association.

Originally published as: *"Darwins Bilder"*
Copyright © 2007 Fischer Taschenbuch in der S. Fischer Verlag GmbH, Frankfurt am Main
Copyright © 2010 by Yale University.

Designed by James J. Johnson and set in New Aster type by Westchester Book Group, Danbury, Connecticut.
Printed in the United States of America by Sheridan Books, Ann Arbor, Michigan.

Library of Congress Cataloging-in-Publication Data

Voss, Julia. 1974–
    [Darwins Bilder. English]
    Darwin's pictures : views of evolutionary theory, 1837–1874 / Julia Voss ; translated by Lori Lantz.
        p.   cm.
    Includes bibliographical references and index.
    ISBN 978-0-300-14174-0 (hardcover : alk. paper)
    1. Evolution (Biology)   2. Darwin, Charles, 1809–1882.   3. Zoological illustration—History.   I. Title.
    QH366.2.V67413 2010
    576.8—dc22

                                                    2010003829

A catalogue record for this book is available from the British Library.

This paper meets the requirements of ANSI/NISO Z39.48-1992 (Permanence of Paper).

10 9 8 7 6 5 4 3 2 1

CONTENTS

ACKNOWLEDGMENTS

This book would not exist without the support of many people and institutions. I would like to thank Hans-Jörg Rheinberger as well as my colleagues at the Max Planck Insitute, especially Peter Geimer, Michael Hagner, Anke te Heesen, Bernhard Kleeberg, Wolfgang Lefévre, Staffan Müller-Wille, Henning Schmidgen, Skuli Sigurdsson, and Margarete Vöhringer.

I am very grateful to Nick Hopwood for reading the manuscript and for his valuable suggestions, and to Philipp Osten and Cara Schweitzer, who also read the manuscript.

I also owe thanks to Frank Steinheimer for his help related to the ornithological collection of the British Museum and the Natural History Museum in Berlin and for the discussions about the history of ornithology, which never failed to widen my horizons.

I also owe a debt of gratitude to the staff of the Cambridge University Library and the Darwin Correspondence project, especially Paul White, who gave me much advice and constructive criticism.

Thank you to Tori Reeve for her assistance at the Down House archive.

The Max Planck Institute for the History of Science, the Volkswagen Foundation, and the Akademie Schloss Solitude all provided valuable financial or institutional assistance.

In addition, Jochen Büttner, Alexander Damianisch, Max Müller-Härlin, and Kostas Murkudis provided highly appreciated support, corrections, advice, and more.

I thank Barbara Liepert of the *Frankfurter Allgemeine Zeitung* Sunday edition for sending me around the world in the tracks of Darwin and Wallace.

I thank my two anonymous referees for their helpful and supportive criticism. My sincere thanks go to Jean Black, Joseph Calamia, and Jeffrey Schier of Yale University Press for making this book possible and for parenting the project in such an enjoyable way. I thank the Börsenverein des Deutschen Buchhandels for subsiding the translation. Finally, I cordially thank Lori Lantz for her elegant translation.

And last, I would like to express my heartfelt gratitude to my parents.

DARWIN'S PICTURES

INTRODUCTION

IN July 1871 the English satirical magazine *Fun* published a caricature of the theory of evolution and its creator (Figure 1). The man in question—Charles Darwin—was now sixty-two years old and had become famous as the author of one of the most significant books of the nineteenth century. In fact, several of his works are still well known today: *Journal of Researches into the Geology and Natural History of the Various Countries Visited by HMS Beagle,* written at the age of thirty, *On the Origin of Species by Means of Natural Selection, or the Preservation of Favored Races in the Struggle of Life,* published when he was fifty, and *The Descent of Man and Selection in Relation to Sex,* which appeared in 1871 and was the subject of the caricature a few months later.[1]

The picture from *Fun* combines two elements. The first is Charles Darwin himself, squatting and with a handkerchief fluttering from his pants pocket like an animal's tail—an image playing on the idea of mankind's kinship with the apes. The other component is the series of drawings mounted on a blackboard, at which Darwin gestures with a pointer. The eleven numbered drawings illustrate "a little lecture by Professor D-N on the development of the horse." Number 1 opens the series with a picture of horseradish, while number 11 concludes it with one of an English

Figure 1. Caricature from the July 22, 1871, issue of *Fun* magazine (Darwin Archive [DAR 255.183], with the permission of the Cambridge University Library Syndicate)

racehorse. The visual joke caricaturing Darwin's theory as positing the evolution of racehorses from vegetables is accompanied by a pun in which horseradish, by virtue of its name, finds a place in the evolution of the horse.[2]

As a historical document parodying actual events, the caricature falls short. From the thorough work of Darwin's biographers we know, for example, that he never gave a single public lecture. Publicity-shy and plagued by chronic stomach problems, he avoided speaking in public his entire life. Photographs or engravings do exist, however, that show both his ally Thomas Henry Huxley and his opponent Richard Owen standing in lecture halls and pointing at blackboards. Darwin, in turn, never taught at a university or spoke before a large audience. Furthermore, the historical facts do not correspond to the contents of the lecture Darwin is shown presenting; the evolution of horses was not a topic he explored. Instead, it was Huxley and Owen who lectured on the fossil remains of the horse family, whose origins date to the Paleolithic era. Huxley spoke about the horse fossils to an "overflowing lecture hall" at the Royal Institution in April 1870, claiming that they offered proof of the existence of evolution. Owen, in turn, had already published on the topic in the 1840s, asserting that the phenomenon proved his "archetype theory," in which everything in the animal world was based on one of four structural plans.[3]

Yet, despite these errors, the caricature draws our attention to an important characteristic of evolutionary theory: the high level of visibility it enjoyed during the nineteenth century. Darwin as well as his theory were familiar enough to be parodied in a popular picture, a situation which is quite unusual in the history of science. Very few scientists or scientific images enjoy the prominence granted to politicians, actors, or famous works of art—all more common subjects of caricature. But scholarly literature has shown how closely Darwin is identified with his work, as the catchword "Darwinism" suggests. The worldview based in evolutionary teachings could just as easily have been called "Wallacism" after Alfred Russel Wallace, who formulated the theory of evolution at the same time as Darwin but never became as famous. It was Darwin, the bearded naturalist of Downe, who became the personification of evolution.[4] And Darwin's theories, as we will see, are closely linked with particular modes of visual representation.

When this book was originally published in Germany three years ago, little was known of the visual history of evolutionary theory. In a pioneering effort, art historian Phillip Prodger had explored Darwin's photographic archive and the scientist's close collaboration with photographers as he studied the expression of the emotions in man and animals.[5] Janet Browne, the renowned biographer of Darwin, had documented the astounding speed with which images of evolution entered popular culture after the appearance of Darwin's *Origin of Species* in 1859. Bowdlerized depictions of evolutionary theory quickly appeared in the form of satirical publications, children's books, china figurines, company logos, and product packaging.[6]

Exhibitions marking the 150th anniversary of the publication of *Origin of Species* in 2009 further expanded our awareness of the visual culture that surrounds evolutionary theory in general and Darwin's writings in particular. The "Darwin effect," a term coined by American art historian Linda Nochlin to describe the immediate impact of Darwin's theory on certain artists, has turned out to be an even broader phenomenon than Nochlin suspected. From the most prominent proponents of impressionism and symbolism to less celebrated but commercially successful and widely known visual artists—the ramifications of evolutionary thought could be seen everywhere. Exhibitions such as *Darwin and the Search for Origins* in Frankfurt and *Endless Forms: Charles Darwin, Natural Science and the Visual Arts* in New Haven and Cambridge demonstrated the Darwinian component in Western visual culture at large.[7]

While understanding how evolutionary theory has changed our way of seeing and interpreting the world is important, this book takes a different approach. The pictures that it explores do not primarily represent the reception of Darwin's ideas, but rather are visualizations that helped Darwin to formulate his theories in the first place. A glimpse into the Darwin archive at the Cambridge University Library suggests the importance that pictures must have held in Darwin's research. The binders, boxes, books, and magazines contain an extensive and diverse collection of visual material and show the care with which Darwin amassed pictures. Even today the archive preserves a vast number of studio photographs of men, women, and children; medical and anthropological shots; and copper etchings, woodcuts, and lithographs of exotic and common animals. They are supplemented by drawings, elaborately colored in some cases, sent by his correspondents from all over the world. Letters accompany them or are even simply worked into the drawings themselves in miniature. In one such letter, two butterfly wings that a correspondent in the Brazilian rainforest pasted to a piece of paper and then mailed bears witness to the visual nature of natural history. This fragile cargo traveled to Darwin in England and has survived into the twenty-first century in its original envelope, packed away in a gray carton.[8]

The books and scholarly journals that Darwin read and kept in his library also contain many illustrations. Furthermore, he often drew himself—from making sketches in his early notebooks to correcting the illustrations prepared for his books. The number of pictures in his books on evolution increases from volume to volume: while the first edition of *Origin of Species* from 1859 contains just a single illustration, this number climbs to seventy-eight in the 1871 *Descent of Man*. *The Expression of the Emotions in Man and Animals* from 1872 features not only twenty woodcuts, but also seven plates comprising between two and seven photos each. Darwin watched over all these illustrations with a careful eye. He either drew them himself, selected them from his collection, or gave precise instructions for their design to the artists commissioned to prepare them. He insisted inclusion of a number of illustrations, often ov
of his London publisher, John Murray, who c

they drove up costs and reduced the profit margin on Darwin's books. When necessary, Darwin himself swallowed this loss by paying for the preparation of the plates himself. He even kept copies of caricatures like the one that appeared in *Fun*, although this material constitutes only a small part of the collection as a whole.[9]

Darwin's visual collection continued to grow over his lifetime of research. Despite the considerable size of the archive, however, it is no longer complete. Letters from Darwin's correspondents often refer to originally enclosed photos or engravings that are now lost. As a result, researchers must sometimes reconstruct the location and context of pictures to which Darwin or his correspondents refer. In other cases, these images can still be found in the archive as a reproduction or original drawing, many of which are published here for the first time.

At the heart of this book are four of Darwin's images that provide a chronological journey through his work. Researchers still primarily published in book form in Darwin's time (only later did the article became the more common publication venue), and each of these illustrations comes from one of his books. Chapter 1 examines the picture of the Galápagos finches that Darwin added to the second edition of *Journal of Researches* in 1845 (see Figure 2). Chapter 2 is devoted to the evolutionary diagram that served as the only illustration to *The Origin of Species* in 1859 (see Figure 33). Chapter 3 looks at a series of pictures showing various stages in the evolution of the pattern found on the wings of the Malaysian argus pheasant—a series of the type parodied in the *Fun* caricature (see Figure 41). Finally, Chapter 4 focuses on the picture of a laughing monkey (see Figure 78) that appears in *Expression of Emotions*, which was published in 1872. These publication dates of 1845, 1859, 1871, and 1872 provide a framework for referring to the rest of Darwin's oeuvre. The chronological sequence also makes it possible to link the pictures to important events in Darwin's life, such as his journey around the world, his return to London, or his move to Downe in the county of Kent, an hour from London. The discussion here follows the order in which the pictures were published, even when this sequence deviates from that of their creation. Back in 1837 Darwin had written the words "I think" in a notebook and sketched a diagram beneath them—his first picture of the evolutionary process (see Figure 12). However, because the image of the Galápagos finches was published in 1845 and the diagram, despite being drawn earlier, did not appear until 1859, the bird illustration forms the topic of the first chapter.

The notebook sketch from 1837 marks the point at which Darwin began producing images of evolution. This particular current of his work ended in 1874 with the appearance of the expanded second edition of *Descent of Man*. This revised version of the book contained many new illustrations, including the series of pictures examining the argus pheasant's showy plumage. The works that followed included numerous illustrations, but they are not, strictly speaking, images of evolution in the sense that they do not systematically explore Darwin's theory. For this reason, this book does not discuss Darwin's work after 1874.

Finally, these four pictures were selected with the present day in mind: the four types of images they represent are the ones biology textbooks, books about evolutionary theory, posters, or natural history museum displays use to explain evolution. They have become icons that every schoolchild recognizes—so omnipresent that mass media, such as advertising, can play off references to them.[10] And as the caricature from *Fun* shows, this phenomenon could already be observed early in the history of the theory of evolution. Faced with this situation, the historian must ask, Why are images and evolutionary teachings so closely entwined?

Answering this question involves recounting the story of how evolutionary theory came to be, tracing the way the Darwin developed his ideas in the nineteenth century by tirelessly creating, reworking, and revising pictures. Although scientists before Darwin had formulated related theories or even evolutionary concepts, his views were a radical departure from those of his predecessors. These breaks with tradition can be followed step by step in his sketches. Two main aspects of his thinking were new: his theory that, if inheritable, tiny differences among organisms form the basis of species change, and the creative role he assigned to extinction. Thanks to the many fossil discoveries during this time, every layperson in the nineteenth century knew that species could become extinct; and many naturalists had already determined that species compete and can eliminate others by, for example, eating them. But it remained for Darwin to incorporate extinction and survival into a system that examined the potential advantages or disadvantages of differing characteristics, as well as the theory that animals select their mates with an eye to variation.[11]

What images did Darwin himself have in mind when he initially sketched out his theory in a diagram? In his 2005 book *Darwins Korallen* (Darwin's Corals), art historian Horst Bredekamp places Darwin's ideas about evolution within the tradition of art collecting as practiced since the sixteenth century. This book arrives at a different conclusion: that Darwin's diagrams are responses to a specifically nineteenth-century approach to collections. These new collecting practices also help explain why Darwin's theory of evolution differs from those of his predecessors.

England's rise to become the largest global colonial and trading power enabled London's British Museum to acquire the most comprehensive natural history collection in the world, and this situation had a number of consequences.[12] The national museum differed from aristocratic art collections in several ways: works of art and natural objects were displayed separately, for example, and the collection was the property not of a private person, but of the nation. Most important, while princely collections had focused on the rare and unique, the bulk of the material in the new museum consisted of objects that served to typify a particular animal or artifact. In the name of the state, explorers, traders, and scientists—including Darwin during the journey of the *Beagle*—combed through Africa, Asia, Australia, and

South America, as far as the distant Galápagos islands, and brought back specimens in untold quantities. More was collected than could be processed in decades, perhaps even centuries. Even today the Natural History Museum in London, which took over the natural history collection of the British Museum in 1881, houses unopened boxes dating from this period of unbridled collecting fever. Unlike earlier collections, which might contain a single particularly beautiful or rare animal specimen, the museum had hundreds or even thousands of such items. Because of their overwhelming number, for the first time one could recognize the room for variation within species. This disorder-producing overcollecting, and the museums' resulting loss of control over their own content, led Darwin—and Wallace—to ponder evolution. That these two scientists developed the same theory at the same place and time supports the idea that the theory of evolution and the images associated with it sprang from factors located in nineteenth-century England. What remains to be shown is how this new image of nature initially drew on the dense symbolic system that had developed in natural history in the waning years of the previous century to depict considerations such as time, variety, or change.[13]

In this book, "Darwin's pictures" can mean one of three things: pictures that Darwin drew himself, those that served him as models, or those that he commissioned from photographers, draftsmen, engravers, or lithographers. In terms of Darwin's own skills, it is necessary to clarify initially that Darwin was not a good artist, at least in the classical sense. Between 1818 and his graduation in 1825 at the age of sixteen, he would have received drawing instruction at the boarding school he attended in his hometown of Shrewsbury. Drawing also would have been a minor part of his studies in natural history in Cambridge and Edinburgh, where he also learned the other craft of the naturalist, preparing animal specimens, from a former slave who had escaped from America. However, when it came to the ability to accurately depict animals, plants, or anatomical details, training was not enough. As a result, as Darwin wrote in his autobiography, "from not being able to draw . . . a great pile of MS. which I made during the voyage has proved almost useless."

Near the end of Darwin's life, the painter and art theorist John Collier, who painted Darwin's portrait in 1881, gave the scientist drawing lessons. Darwin thanked him and once again complained that he "could never draw a line."[14] By that time, however, he had learned to have others draw for him when his own skills were insufficient, or to collect the pictures that came to form an archive of his observations. The forgotten craftsmen who worked under his direction play a recurring role in this book.

By presenting Darwin as a virtuoso artist, the *Fun* caricature is thus again inaccurate. As previously noted, it was Huxley and Owen who could draw brilliantly and who put this talent to use in their lectures, creating pictures on the board before the eyes of the audience. Ernst Haeckel, the famous German champion of evolutionary theory whose books such as *Art Forms in Nature* influenced the art deco movement, also springs to mind.[15] However, objective drawing from nature is only one aspect of this skill; an area in which Darwin had significantly more practice and achieved a certain level of mastery was geological illustration (Figure 26 [Color Plate 5]). At seventeen he began his medical studies in Edinburgh and attended Robert Jameson's geology classes, where he learned to identify the accumulations of stone material in the ground as layers and depict them in cross-section. The ability to identify a stone with a particular layer of the earth, to connect a fossil to a geological epoch, or to translate the earth's hidden interior into large-scale drawings or colorful maps all were things he continued to practice aboard the *Beagle*, as entries in his notebooks and diagrams from this time show.[16]

With this work he learned early to understand periods of time that are beyond our ability to grasp directly and bring them to life them in pictures. By the end of the eighteenth century, at least two disciplines were confronted with the problem of how to register processes that unfold so slowly that humans cannot observe them—or that even take longer than the human life span. New kinds of images were the result. In the decades before Darwin, embryologists learned how to present the development of organisms in series of individual pictures. During the same period, geologists portrayed the layers in the earth's interior and their ages in the form of cross-sectional diagrams. This translation of form and time into a symbolic system of rows, lines, angles, or points opened up space for Darwin to conceptualize his theory of evolution. It taught him to think in terms of millennia, to consider the impact small changes in nature could have over long periods, and to imagine how a line might continue based on its angle of inclination. The fact that he dedicated the second edition of his travel book to Charles Lyell, his lifelong supporter and the most important English geologist of his day, demonstrates this proximity. By giving observers access to conceptual realms that were inconceivable previously, the images considered in the following chapters fulfill the function Paul Klee claimed for twentieth-century art: they do not reproduce the visible but make visible.[17]

Another peculiarity in Darwin's view of nature becomes evident in a letter that he sent to his friend Joseph Dalton Hooker, a botanist, in 1856. After musing about the mouth parts of mollusks, Darwin makes a most remarkable statement, claiming "What a book a Devil's chaplain might write on the clumsy, wasteful, blundering low & horridly cruel works of nature!" Of course, Darwin himself went on to publish this very book in 1859. Both the words and the images in *Origin of Species* describe the two principles of evolution—the "blundering" and the "cruel works of nature" we know better as random change and natural selection. But while the written account and its plot of "the struggle for survival" emphasize selection, the pictures are dominated by the principle of randomness. Darwin expressed the ideas of competition, effort, and struggle in the medium of language, culminating in the expression "survival of the fittest," which has taken on a life of its own and has become a byword in many languages.[18] The other evolutionary principle, however— that of chance, variation, disorder, and incompleteness—was refined in images.[19] Selection appears in the pictures only in terms of its consequences: when a line in an evolutionary diagram abruptly ends, the species it represents is extinct. The notion of chance, in turn, drives the very structure of this image in the form of the lines radiating from the nodal points that symbolize random variation.

His insistent focus on the imperfections, deficiencies, and peculiarities of living things provided Darwin with the nineteenth century's most unusual perspective on nature. He peered into the cracks and tears of what had seemed liked God's perfect creation until these gaps widened into a door to evolutionary theory. The "complete millionaire in odd & curious little facts," as he once called himself, noticed and recorded phenomena that his colleagues, who viewed nature as a work of art, failed to notice—even after his works became famous.[20]

Darwin's pictures were produced with simple means: pencil, paper, the naked eye. In most cases they were reproduced using woodcuts, the most common technique in the nineteenth century. In one book, however, Darwin took advantage of both a new recording technology and a new printing process: *Expression of Emotions* employs photography and the heliotype method. But regardless of the tools used, Darwin's letters and the instructions he sent to draftsmen or printers show the high degree of artistic precision required to depict chance on paper. To his frustration, the qualities of symmetry, order, and regularity tended to emerge more often in pictures than they do in nature itself. Whenever the specialists charged with producing his images smoothed out or "improved" his drafts, he sent them back with corrections, whether the pictures were abstract diagrams or representational images depicting apes, peacocks, humans, or other creatures. Working out his theory in images involved constant struggle over millimeters of difference—from the first drafts to the final plates.

The following pages may contain aspects that the title *Darwin's Pictures* did not lead readers to anticipate and omit subjects they were sure would be covered. Those familiar with Sternberger's description of evolutionary theory as "an imposing painting of nature's eternal warfare"—or who recall the battles shown in countless books about prehistory and dinosaurs from their youth—may expect more of the same from Darwin. They will be disappointed.[21] It is true that the interpretation of selection in nature as necessarily violent—a view that implicitly equates evolutionary theory with so-called social Darwinism—dominates visual responses to Darwin's work in both the arts and popular culture. However, Darwin's own pictures—the ones examined in this book—are of a different nature. They highlight the other principle of evolutionary theory: variation.[22]

Similarly, some readers may be bracing for racist images of the kind circulated well into the twentieth century that suggest some groups of people are closer to the animals than others. This foreboding will also go unfulfilled: Darwin never produced such pictures.[23] Others did respond to the idea of evolution by generating these kinds of images, but in a history of how Darwin's theory emerged they can be discussed only in isolation. No matter how valuable a study of racist misinterpretations of evolution would be, they are not the subject of this book.

The idiosyncrasy of Darwin's approach to illustrating his books can be demonstrated by leafing through *Descent of Man*. Despite its title, the book does not include a single picture of a human being. Such surprises are not uncommon in Darwin's work, and they can provide keys to understanding the theory of evolution. Finally, despite a common belief to the contrary, Darwin's pictures hardly make the world a less magical or mysterious place. The lovely words of writer Osip Mandelstam testify to their heartening effect: "The inspiring clarity, like a clear day in a moderate English summer, the—if I may call it this—good weather for science, and the author's restrainedly elevated mood rub off on the reader and help him to make Darwin's theory his own."[24]

1 THE GALÁPAGOS FINCHES

John Gould, Darwin's Invisible Craftsman,
and the Visual Discipline of Ornithology

In October 1836, the research ship H.M.S. *Beagle* returned to the port of Falmouth, England. It had sailed the oceans for five years at the behest of the world's largest colonial power, Great Britain, visiting South America, the Galápagos islands, Tahiti, New Zealand, Australia, Mauritius, and Cape Town. The captain and his crew had spent more than half of this time mapping the coast of South America as well as filling in existing maps more precisely. Outfitted with the most up-to-date technical equipment available, the *Beagle* had the primary task of determining the exact longitude of the Brazilian town of Bahia—today Salvador—which was indicated differently on French and German maps. During its journey the *Beagle* crossed paths with a number of other vessels sailing along the South American coast under the British flag. From 1831 to 1835, two hundred and fifty British trading ships sailed to South America to drop off goods and purchase raw materials. Furthermore, two additional research expeditions set off shortly after the *Beagle* sailed and had the same charge of creating maps, exploring the interior, and keeping an eye out for sources of precious metal and stones.[1]

This trip was the second for the *Beagle*—a fresh attempt to complete a voyage of discovery after the first had ended in tragedy several years earlier. The ship's first captain shot himself while the boat was moored in Tierra del Fuego, thousands of miles from his English homeland, leaving behind a crew suffering from scurvy and a labyrinth of erroneous maps and drawings. His successor was the young Captain Robert FitzRoy, who first brought the ship safely back to England and later led her on two expeditions of his own. The *Beagle*'s second journey proved to be a success, resulting in not only new maps but a valuable collection of animals, plants, minerals, and fossils from the places it had visited. The *Beagle* was a rather small ship—a two-master just ninety-eight feet long and about twenty-six feet wide—and little space remained for the collection after accounting for the needs of its seventy-four-man crew. Many crates were shipped back while the ship was still under way, and just a few actually reached England in the hold of the *Beagle* itself. News of the ship's amazing discoveries in South America thus sped back to England ahead of the *Beagle*'s arrival. For example, the fossil skull of a giant sloth, found in Argentina in 1832 and immediately shipped back to England, was displayed at a meeting of the British Association for the Advancement of Science (BAAS) in Cambridge in the summer of 1833, three years before the ship's return.[2]

But even without sensational discoveries, an expedition to South America was sure to arouse public interest. Ever since Alexander von Humboldt's *Travels to the Equinoctical Regions of America* and *Views of Nature* had appeared, travelers to South America could count on a broad following. The continent was considered a paradise for naturalists, and its tropical regions were synonymous with exotic adventure.[3] Like moths to a flame, generations of researchers were lured to the distant lands portrayed in Humboldt's texts and pictures at the beginning of the nineteenth century by the romantic enthusiasm that infused them. Those who remained at home, both experts and laymen, longingly waited for the latest news of the teeming beauties of Brazil, Argentina, Chile, or Patagonia. "At present I am only fit to read Humboldt; he like another Sun illumines everything I behold," noted the *Beagle*'s most famous passenger, the young

Cambridge graduate Charles Darwin, as he described the impression Humboldt's writings had made on him. Eight months before the ship set sail he wrote to his sister Caroline: "In the morning I go and gaze at Palm trees in the hot-house and come home and read Humboldt: my enthusiasm is so great that I cannot hardly sit still." He would be able to relax, he added, only when he saw Humboldt's great dragon tree with his own eyes.[4] At the same time, he prepared himself for his journey by studying pictures such as the lithographs of tropical forest scenes by the German painter Johann Moritz Rugendas. He would recall these scenes when he first set foot in the real jungle in 1832.[5] The young Darwin had yet to leave English soil when he was sweating out his tropical fever in the hothouses of Cambridge and dreaming of discoveries and adventures to come. And although he would end up traveling the world, he would never set foot on the Continent: Darwin visited Rio de Janeiro, Montevideo, St. Helena, and Sydney, but he never saw Rome, Paris, or Berlin.

There was never any question that the *Beagle* would encounter previously unknown species of animals and plants. As a direct consequence of the increase in global trade, the number of newly discovered species grew faster in the first half of the nineteenth century than ever before. The world's waterways pulsed with traffic between England and its colonies, and the ports filled up with larger and larger ships—ships that carried not only silk, rubber, spices, coffee, ivory, gold, and silver over the seas, but also living things.[6] Business was booming on London's docks for animal dealers such as Charles Jamrach, who bought the exotic cargo, dead or alive, to supply museums, zoos, exhibitors, and menageries throughout Europe. The overwhelming wealth of new discoveries meant that universalists in the mold of Carl Linnaeus became a rare breed. Few scientists stepped up to succeed the eighteenth century's great Swedish naturalist, who had devised a scheme to classify not only all known animals and plants, but minerals as well.[7] Specialists devoted to every division of the animal and vegetable kingdoms emerged, and as a result new species were increasingly unlikely to be discovered during an expedition itself. Instead, they were first identified back at home by scientists working in museums. In the field, thousands of miles from the closest zoological collection, it was impossible to compare an animal with similar specimens. Without such comparison, it was difficult to determine whether a find truly represented a new species or one that already had been identified.

Very few beetles, fish, frogs, birds, or mammals are so spectacularly unique that researchers in the 1830s could immediately tell whether a specimen they encountered had previously been seen or collected on one of the many earlier journeys. Even sensational finds were often first identified only after they had been delivered to a museum and turned over to the specialists there. While the young Darwin prepared twelve catalogs during the journey as a record of the plants and animals collected, these were just preliminary. Furthermore, the twenty-two-year-old, who had a degree in theology, basic knowledge of medicine, and a great interest in geology, had been invited to join the trip as a companion for the captain: to avoid falling prey to the loneliness that had driven his predecessor to suicide, Fitz-Roy had requested the company of a gentleman with standing and education to match his own. Once the voyage was under way, however, Darwin took on the function of ship's naturalist as well.[8]

An enthusiastic collector since childhood, and equipped by his studies with a knowledge of scientific collecting, Darwin conscientiously assigned every object found during the voyage to a taxonomic category and gave it a preliminary name. But the task of accurately determining the species to which each specimen belonged was well beyond the expertise of any single zoologist, especially one so inexperienced. Darwin's task as ship's naturalist was therefore less to identify the species of the finds than it was to organize, record, and store them. At sea he sorted what was found in the nets and oversaw the preservation of selected discoveries. On land he personally made long expeditions or, when he could not go himself, gave his servants extensive instructions on which kinds of animals to hunt, collect, and preserve. He personally tied a paper label to every specimen brought on board and marked it with a number that corresponded to an entry in the catalog, which typically contained the name and sex of the object, as well as the date and location of its discovery. However, in a number of cases Darwin failed to indicate the location, and he wrote the names inconsistently in English, Latin, or the local language—usually Spanish. The ornithological notes that use Latin names are particularly prone to errors.[9]

Darwin's five years at sea provided the fodder for ten publications between the years 1838 and 1846: *The Zoology of the Voyage of the H.M.S. Beagle* alone consisted of five volumes, and *The Geology of the H.M.S. Beagle* included three; these were accompanied by the *Journal of Researches*, which appeared in 1845 in a revised second edition. Later Darwin also produced a four-volume study of barnacles, *Monograph of Cirripedia*, which he published from 1851 to 1854.[10]

Evolution is never mentioned in any of these fourteen volumes. Although his notebooks show that he had speculated about evolution since 1837 and began planning to publish his theory in 1842, he avoided mentioning the subject until 1859, with the appearance of *On the Origin of Species*. For two decades his readers saw no sign of the provocative theorist he later proved to be, but knew him rather as an explorer, geologist, and expert on mollusks whose solid work had been distinguished with the Royal Medal. This specialist knowledge helps explain the acceptance evolutionary theory later enjoyed—even his opponents could not claim that he did not know his field.[11]

The history of Darwin's theory of evolution—or at least its public history—thus seems to begin in 1859. But such a view overlooks an exceptional instance in which Darwin, for a single moment, broke this twenty-year silence. Without ever using the word "evolution," Darwin explained his theory of species change to his readers as early as 1845, in the second edition of the *Journal of Researches*. This anticipatory comment was the first instance of the foreshadowing which marks much of his work. In 1859, for example, the statement "light will be thrown on the origin of man and his history" appears in *Origin of Species,* the only mention of human evolution in a book that otherwise scrupulously avoids the subject.[12] It was twelve years later, with the 1871 publication of *Descent of Man,* that Darwin first openly admitted that he considered humans to be products of natural history just as animals are. The anticipatory maneuver from 1845, fourteen years before his first publication on evolutionary theory, is even more remarkable. Darwin's research on human evolution after 1859 was presaged by his memorable remark in *Origin of Species,* but his line of thought grew from the ramifications of the entire book. In 1845, in contrast, a parenthetical discussion of barely half a page turned a remote and unremarkable object into a subject of evolutionary inquiry. Without Darwin, this creature—the Galápagos finch—would likely never have interested researchers in the field. Yet the inconspicuous gray-brown birds became one of the most famous animal populations in the world—the exemplary subject of evolutionary research and one still studied today by the members of the Charles Darwin Research Station, which was founded on the Galápagos in 1959. The finches enjoy this lavish scientific attention despite the fact that Darwin—as historians often point out—mentions them only in this isolated passage, never returning to discuss them in his writings on evolutionary theory from 1859 or later.[13]

Darwin's reference to the birds in *Journal of Researches* in 1845 is, strictly speaking, not addressed to the reader but to the viewer, as it occurs in an illustration specially added for the book's second edition (Figure 2). If we open this edition of Darwin's travel book to the seventeenth chapter, we discover depictions of four of the thirteen species of finches collected during the *Beagle's* journey, compressed into a picture that takes up half a page. Only their heads are shown, turned in profile to the left and treated identically in terms of cropping, color, and size. This comparative perspective clearly delineates the deviations in the shapes of the birds' heads and beaks.

At first glance the four finches look very different from one another, although they all belong to a single genus. This picture

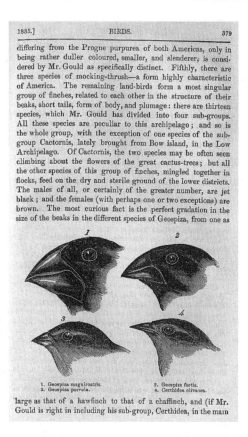

1835.]   BIRDS.   379

differing from the Progne purpurea of both Americas, only in being rather duller coloured, smaller, and slenderer, is considered by Mr. Gould as specifically distinct. Fifthly, there are three species of mocking-thrush—a form highly characteristic of America. The remaining land-birds form a most singular group of finches, related to each other in the structure of their beaks, short tails, form of body, and plumage: there are thirteen species, which Mr. Gould has divided into four sub-groups. All these species are peculiar to this archipelago; and so is the whole group, with the exception of one species of the subgroup Cactornis, lately brought from Bow island, in the Low Archipelago. Of Cactornis, the two species may be often seen climbing about the flowers of the great cactus-trees; but all the other species of this group of finches, mingled together in flocks, feed on the dry and sterile ground of the lower districts. The males of all, or certainly of the greater number, are jet black; and the females (with perhaps one or two exceptions) are brown. The most curious fact is the perfect gradation in the size of the beaks in the different species of Geospiza, from one as

1. Geospiza magnirostris.   2. Geospiza fortis.
3. Geospiza parvula.   4. Certhidea olivacea.

large as that of a hawfinch to that of a chaffinch, and (if Mr. Gould is right in including his sub-group, Certhidea, in the main

Figure 2. The Galápagos finches in the second edition of Darwin's *Journal of Researches,* from 1845

presents us with the morphological diversity a genus can encompass, but it also does more. The composition encourages us to read the picture from the upper left to the lower right, and along this suggested path the birds are arranged from the largest, heaviest type to the smallest and most delicate. The differences between the species thus appear to be the products of continuous change—a process occurring in time. This critical arrangement becomes clear if we try switching the locations of the first and third finch: the resulting picture still demonstrates the variety within a genus but implies nothing about the relationship among these differences. The composition arranges what would otherwise be mere differences into a pattern. The order imposed by Darwin allows each head to serve both as a representative of its own species and as a potential predecessor to those that follow. Darwin also referred to the new illustration in the revised text, writing that "seeing this gradation and diversity of structure in one small, intimately related group of birds, one might really fancy that from an original paucity of birds in this archipelago, one species had been taken and modified for different ends."[14]

In his unpublished drawings Darwin was already championing the process he describes here merely as something "one might fancy," and he would later describe it in more detail in *Origin of Species* using an example based on pigeons.[15] He declares that the "tendency to vary" exhibited by finches, pigeons, and other organisms is a prerequisite for species change; continuous deviation among organisms allows varieties to become new species, species to become new genera, and genera to become new families through the course of evolutionary history.[16] The tiny idiosyncrasies that distinguish one organism from another can, if inherited over generations, cause a new species to develop from what was originally just an intraspecies variation. When Darwin used the finches to illustrate his theory of species splitting in 1845, they represented the first visualization of the theory of evolution that he presented to the public. Although this solitary image was buried in the seventeenth chapter of a second edition, it became an icon. The birds have since been discussed in countless textbooks as an example of how species split off, and their image appears on stamps, posters, and T-shirts.[17] The power of the 1845 illustration exemplifies how images can formulate and convey scientific theories, and it offers a lesson in how pictures were used to communicate knowledge in the nineteenth century.

One aspect of the untold visual history of the Galápagos finches forms one of science's most enduring myths. The illustration creates the impression that the theory of evolution occurred to Darwin during the voyage itself, as if the scales fell from his eyes during his five-week stay on the Galápagos and the picture is a record of this event. As scientific historian Frank Sulloway has shown, however, what really happened was quite different. The tale of Darwin's supposed "Eureka!" moment not only distorts how he arrived at the theory of evolution, but also obscures the real discoverer of the Galápagos finches: the bird illustrator, ornithologist, and publisher John Gould. Gould was Darwin's "invisible craftsman." In his own time, however, Gould was anything but invisible: he held a prominent position in the lively universe of ornithological artists, embalmers, and museum curators that made Darwin's finch picture—and as the insights it inspired—possible.[18]

ZOOLOGICAL COLLECTIONS AND PICTURES IN LONDON

When Darwin returned to England in early October 1836, his first order of business was to decide how to divide the expedition's crates and bottles full of skeletons, pelts, skins, organs, fossils, beetles, worms, and fish among the specialists at the capital's zoological societies and institutions. Three potential recipients for the collection were in London. The first was the British Museum, England's famous national institution. Founded in 1754, the British Museum held not only natural history specimens, but books and antiquities as well. The second, only a few minute's walk from the British Museum on Russell Street, was the Hunterian Museum in Lincoln's Inn Fields. Its collection from the Royal College of Surgeons provided an opportunity for physicians- and naturalists-in-training to study the comparative anatomy of vertebrates. The third museum was the newest of the three but it had made a good name for itself in just a few short years: the Zoological Society, located in the southwest section of the British Museum, maintained not only a collection of preserved specimens but also a constantly growing group of live animals kept in the Zoological Garden in Regent's Park. Compared with the British and Hunterian Museums, the Zoological

Society was not just new, but liberal. It accepted female members, something that other London organizations such as the Linnean Society considered unthinkable well into the twentieth century. Alongside the Zoological Garden in Regent's Park, which quickly become one of the city's most popular attractions, the society organized exhibitions of its holdings to encourage public interest in zoology and generate additional income.[19]

Before his departure, Darwin obtained the right to freely decide which institution would receive the collections that would accumulate during the trip.[20] His request for this authority was unusual, since until that time an expedition sailing for the English crown would turn over all its treasures to the British Museum automatically. FitzRoy, the *Beagle*'s captain and also a collector of animals and plants during the journey, accordingly delivered all his specimens to that institution. Darwin, in turn, divided his collections among specialists at all three organizations, a tactic which promised quicker viewing and processing of the material. The fossils went to Richard Owen, the comparative anatomist at the Hunterian Museum; the fish and reptiles ended up with Leonard Jenyns at the British Museum; and the birds—four hundred and fifty in all—became the wards of John Gould, curator of the Zoological Society.[21]

For a young institution like the Zoological Society, the *Beagle*'s bird collection offered an opportunity to prove its scientific standing among the other institutions. In the nineteenth century, an era that celebrated discovery and display, obtaining just a single exotic animal could bring lasting fame. The Zoological Society itself demonstrated this phenomenon in 1827, the year of its founding, when a giraffe arrived in London from Mehmet Ali, the viceroy of Egypt and Ottoman pasha. King George IV, notorious for his opulent taste, was immediately taken with the rare and capricious animal, and an eager public followed the adventures of the giraffe at court in newspaper articles and caricatures. Thanks to its distinctive markings, the long-legged animal earned the nickname "camelopard," suggesting a mixture of a spotted leopard and a camel.

The besotted monarch built a special house for the giraffe at Windsor Palace and had its oil portrait painted twice.[22] Guests who had a chance to view the animal were regaled by lines of poetry celebrating its beauty, in which its eyes were compared with those "of the finest Arab horse and the loveliest southern girl" and their expression with "volcanic fire." It was as if a princess from the *Thousand and One Nights* had set up her tent next to the palace. Two years later, the giraffe was dead; it had never fully recovered from the strain of the journey. The task of preserving the royal animal fell to the Zoological Society, and the curator entrusted with the job was John Gould. Gould had obtained his position by winning a taxidermy contest, but the giraffe was far and away the largest animal on which he had ever plied his craft. However, he met the challenge and fulfilled his charge to the king's express satisfaction. The death of the giraffe thus ended in a triumph of taxidermy, and the animal became the crown jewel of the Zoological Society's collection. Gould's work earned the praise of critics as well.[23]

Like the giraffe commission almost ten years earlier, obtaining the birds from the *Beagle* collection was a stroke of luck for the Zoological Society. Many new species were expected to come to light as the specimens were identified—discoveries which could then be reported in the journals *Transactions of the Zoological Society* and *Proceedings of the Zoological Society*. Furthermore, the birds were the most attractive part of the collection.[24] Compared with the fossils, which required a great deal of preparation to arouse broad interest, or the reptiles, which appealed largely to specialists, the birds were sure to attract the public's attention. There was barely a type of animal with so many fans as birds—they fascinated scientists and laypersons, aristocrats and the middle class alike. The large audience of bird lovers inspired ornithological authors to extravagant feats outstripping those in any other field of study. The results were books like *Birds of America*—one of the world's largest books even today—which was published and illustrated between 1828 and 1837 by John James Audubon. On pages measuring almost twenty-seven by forty inches, the viewer could leaf past colorful parrots and bronze-colored eagles to pink flamingos and snow-white swans, all depicted in life size.

In addition to the remarkable format and spectrum of colors, the book offered dramatic illustrations of an entirely new type. With Audubon, ornithology left the contemplative protestant parsonages in which it was born and entered a new, adventurous life.[25] Blood gushes from the eye socket of the hapless rabbit that Audubon's golden eagle carries off in its claws (Figure 3); his mockingbird finds a spitting rattlesnake in its nest and tumbles backward off the branch at the sight of the gaping maw. The drama of life and death enfolding the animals in Audubon's pictures extends to surround his own self-representations. His image, no bigger than a thumb, appears in the background of the golden eagle picture, trying to cross a deep crevasse on a fallen tree. The message is clear: in order to capture nature's image at the dizzying heights of the eagle's aerie, the author was willing to risk his own life. His pictures thus also reveal a point of view shared by many naturalists, whether ornithologists, other zoologists, or botanists. In his 1790 work *Philosophy of Natural History*, for example, the Scottish philosopher and naturalist William Smellie wrote about the cycle of "animation and destruction" in the animal kingdom—a cycle in which each species ensures its own survival through the death of another. He observes, for example, that a pair of sparrows requires 3,360 caterpillars a week to feed their young. Darwin marked both this detail and the passage about the cycle of life and death in his copy of Smellie's book. Audubon's hunting and hunted birds unforgettably depict this aspect of avian life.[26]

To boost sales Audubon marketed his book not only in his home country of America but also in England. In the nineteenth century, England was the world's largest book market, as more ornithological volumes were published here than in the rest of Europe combined.[27] Although no other publication could compete with the size of Audubon's "double elephant portfolio," many had features that were hardly less majestic. "The birds,

Figure 3. The golden eagle from Audubon's *Birds of America*, 1828–1837

for the most part, are life-size," wrote one commentator about the illustrations in a deluxe ornithological edition, "and the tails are as deftly manoeuvred as the ladies' trains at a Drawing-room. So many gorgeous plates of species that are often yet more gorgeous are somewhat overpowering. We turn them over with fear and trembling."[28]

Judging from these ornithological publications, every bird enjoyed an enthusiastic following: local species and exotic ones, powerful predators and charming songbirds, ungainly ducks and tiny hummingbirds. Social status often determined an interest in certain bird types. Country nobility, who enjoyed the right to hunt, turned their attention to pheasants, partridges, and ducks. The urban upper class, in turn, valued birds for their beauty, and the city's parks and gardens filled up with imported peacocks who delighted visitors by unfurling their tail feathers and mingling their hoarse cries with the twittering of their native English cousins. At home, their smaller relatives hopped about in their cages or aviaries and became beloved companions. Queen Victoria considered her birds as worthy of a portrait as she did the other members of the royal family. Her macaw and lovebirds "sat" for Sir Edwin Landseer, the celebrated animal artist, who became famous for the bronze lion cast from a captured French cannon for the Nelson monument in Trafalgar Square (Figure 4). Later, the painter Joseph Wolf would capture the appearance of a royal bullfinch for the ages. Birds in zoological exhibitions also proved to be magnets for visitors. A collection of stuffed hummingbirds that the Zoological Society showed in a pavilion in 1851 is said to have attracted seventy-five thousand spectators.[29]

The enthusiasm for birds gripped not only adults of every educational and income level, but children as well. Charlotte Brontë's *Jane Eyre*, for example, includes a scene in which the young heroine loses herself in the pages of Thomas Bewick's popular *History of British Birds*. The girl escapes from her aunt's maltreatment into the world Bewick presents, sharing nooks in the cliffs with the seabirds seeking shelter from ocean storms. Jane herself explains why she spent so much time perusing Bewick's

Figure 4. Edwin Landseer's 1839 group portrait *Islay, Tilco, a Macaw, and Two Love-Birds* (With the permission of The Royal Collection © 2009 Her Majesty Queen Elizabeth II)

Figure 5. Woodcock from Bewick's *History of British Birds* from 1797

illustrations, stating that "each picture told a story; mysterious often to my undeveloped understanding and imperfect feelings"[30] (Figure 5). When *Jane Eyre* was first published in 1847, fifty years had passed since the appearance of Bewick's *History of British Birds*. The work not only won the hearts of children, but conquered—if not revolutionized—the book market. Bewick's 1797 study of birds introduced an improved woodcut process that allowed text and images to appear on a single plate and thus on the same page. The resulting pages were both of high quality and inexpensive, but the decisive advantage was the fact that illustrations no longer had to be printed on special paper and inserted at the end of the volume.[31] Now that a complex binding process and expensive, high-quality paper were no longer necessary, the number of illustrated books increased considerably. *History of British Birds* thus represents the gateway through which illustrations entered the nineteenth-century English press and book market. The far-reaching consequences of the inexpensive printing process—such as the founding of the first illustrated periodical, *Penny Magazine*—have been described in detail elsewhere.[32] Of course, the specialized field of ornithology, whose history is the focus here, was affected by these changes as well.

The revolutions in printing technology that had made the study of birds an affordable hobby also led to the creation of the first bird identification books—handbooks that have since become indispensable to both professional specialists and hobbyists. Unlike Audubon's luxurious volumes, such books offered simple black-and-white pictures that told the reader less about the beauty of birds than about the features distinguishing one species from another. One of the first efforts of this kind was William Swainson's handbook *The Natural History and Classification of Birds* from 1836. The book was characterized by numerous silhouettes, which Swainson used to portray the species of birds and the characteristics they display (Figure 6). Rejecting any distracting extras and excluding portrayals of the birds' lives and surroundings—the very elements that made Bewick's birds attractive to Jane Eyre—Swainson concentrated exclusively on the birds' bodies, which his indexed illustrations broke down into separate parts.[33] Furthermore, many of his images focused on a single feature—the crest, eye, ears, tongue, beak, wings, or talons. The bird—viewed as the sum of its parts—was abstracted to a two-dimensional shape, and these silhouettes served to heighten the viewer's awareness of detail. The range of beak shapes found within the genera *Tardivola* and *Pitylus*, for example, is illustrated by two rows of bird heads (Figure 7). The coloring and position of every bird in the illustration is the same: all the heads are turned to the viewer's left, the drawings are cut off just behind the each bird's eye, the plumage is always black, and the white beak is marked with a black outline. This standardized uniformity makes the differences among the birds even more apparent—a practice that Darwin would utilize when portraying the Galápagos finches ten years later.

With Swainson, ornithological illustration thus reached the other end of the spectrum: handbooks taught readers a taxonomic awareness of features, while the sumptuously illustrated volumes encouraged readers to marvel at the beauty of the animal kingdom. Both ways of seeing—a standardized, reduced perspective and an open admiration for the wonders of nature—

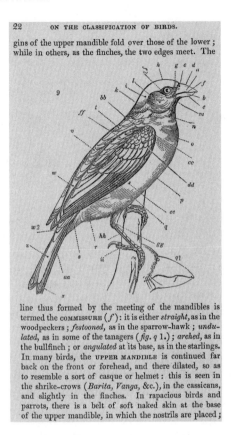

22 ON THE CLASSIFICATION OF BIRDS.

gins of the upper mandible fold over those of the lower ; while in others, as the finches, the two edges meet. The

line thus formed by the meeting of the mandibles is termed the COMMISSURE (f): it is either *straight*, as in the woodpeckers ; *festooned*, as in the sparrow-hawk ; *undulated*, as in some of the tanagers (*fig. q* 1.) ; *arched*, as in the bullfinch ; or *angulated* at its base, as in the starlings. In many birds, the UPPER MANDIBLE is continued far back on the front or forehead, and there dilated, so as to resemble a sort of casque or helmet : this is seen in the shrike-crows (*Barita, Vanga,* &c.), in the cassicans, and slightly in the finches. In rapacious birds and parrots, there is a belt of soft naked skin at the base of the upper mandible, in which the nostrils are placed ;

Figure 6. Illustration from Swainson's 1836 *The Natural History and Classification of Birds*

had their place in ornithological research, although the latter view eventually came to pose an inherent challenge to the theory of evolution.[34]

Thanks to the popularity of bird-watching in the nineteenth century, even the ten-year-old Darwin wondered "why every gentleman did not become an ornithologist." As an adult, he recorded in his notebook the observation that children enjoy looking at pictures of all types of animals—a pleasure that he may have

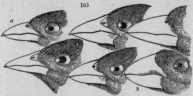

AFFINITY OF TARDIVOLA AND PITYLUS. 115

wonder, then, that some of our best writers, who seem
to be acquainted with one or two of the typical examples of *Tardivola*, should place it as a *genus*, in a
totally different family.   To illustrate this extraordinary union of two subgenera (apparently so widely
separated by characters which are usually considered
indications of higher divisions), we subjoin the outlines
                                163

of the intervening gradations in the form of the bill
just alluded to, and which will bear us out in the belief, that, whatever uncertainty hangs over other parts
of our arrangement of the tanagers, the proximity of
Tardivola to *Pitylus* is beyond dispute.
      (130.) This foregoing affinity being admitted, we
are next to inquire into the cause why such a remarkable variation in the bill should occur in species so
closely united.   Now, it should first be stated, that
nearly the whole of the seed-eating birds of Tropical
America are composed of the tanagers, which, in those
regions, supply the place of the other finches, so abund-
ant in all parts of Europe.   The seeds and hard berries,
however, found in our cold and temperate climates,
are very few indeed, when compared to the innumer-
able variety produced in the vast forests of the New
World, whether we regard the variety of the species, or
the different degrees of hardness they possess.   Now,
as these small and hard fruits are the appointed food of
the tanagers, (for the parrots chiefly subsist upon the
larger nuts,) it follows, that an equal diversity of
strength should be found in the bill; that organ, in
                               ι 2

Figure 7. The shapes of the beaks of Tardivola and Pitylus in Swainson's *The Natural History and Classification of Birds* of 1836

experienced himself. As the son of parents educated in natural
history, he probably made the acquaintance of ornithological
illustrations early in life and may have, like Jane Eyre, perused
Bewick's *History of British Birds* while still a child. During his
studies, he also attended the lectures that John James Audubon
held in Edinburgh. Darwin's copy of Swainson's *The Natural
History and Classification of Birds* is thickly strewn with underlined passages that reveal how closely Darwin read the heavily

illustrated book. Ornithology in the nineteenth century was quite literally a visual discipline and one that, thanks to the numerous expeditions during the period, understood itself as descriptive: its focus was on the diverse outward appearances of birds and the ways to recognize them. Since the shape of an animal—its features and proportions—could be shown more precisely in pictures than with a written description, bird-watching without pictures was as unthinkable as a "zoo without the animals."[35]

In the spring of 1837 Darwin finally met the man who published more ornithological books than anyone else in the nineteenth century. The Zoological Society had made a name for itself with a bird collection even before Gould was commissioned by Darwin to scientifically evaluate the *Beagle*'s birds: a set of specimens from the Himalayas had led to the identification and description of twenty-five previously unknown bird species. This discovery, combined with the rareness of the birds, prompted the Society to publish *A Century of Birds from the Himalaya Mountains*, which was issued in several installments from 1830 to 1831. The work features hand-colored lithographs of a hundred different Himalayan birds, sometimes along with their chicks. The man behind the book—which was highly praised by experts and enjoyed excellent sales—was the same John Gould to whom Darwin entrusted the task of classifying the *Beagle* bird collection in 1837.[36]

GOULD'S CLASSIFICATION SKILLS AND INVISIBLE CRAFTSMANSHIP

The meeting of Gould and Darwin in January 1837 was probably their first. Following his return Darwin spent the first few months at Cambridge, where the *Beagle* collections were stored during and immediately after the ship's voyage. The fossil remains of South American mammals had already been handed over to the Hunterian Museum in December. Richard Owen, the museum's specialist in this area, informed Darwin just two weeks later that the specimens represented a *toxodon* and a *scelidotherium*—or, as Darwin excitedly reported to his cousin: "some gnawing animal, but of the size of a Hippopotamus!" and "an Ant-Eater of the size of a horse!"[37] When Darwin returned from Cambridge to London on January 4, he presented a brief paper on the geology of the Chilean coast, which he had examined during the *Beagle*'s voyage. The files of the Zoological Society record that the Galápagos birds arrived the same day. It seems likely that Darwin took advantage of his visit to London to present the collection to the museum in person. Even before this meeting, however, Darwin and Gould had surely heard of one another.[38] At the time they met, Gould was the more famous of the two.

John Gould, who was born at the royal court in 1804 as the son of a gardener, was only five years older than Darwin, but thanks to his elaborately illustrated bird books he had already achieved a measure of renown. When Darwin set off on his trip around the world in 1831, Gould was issuing the last installment of *A Century of Birds of the Himalaya Mountains*. When Darwin returned five years later, in 1836, Gould had followed up his debut with three additional publications—one on the birds of Europe, and a volume each on the toucan and trogon families of birds. Gould had developed his taxonomic expertise in a roundabout way; his family's simple means did not allow him to enjoy a scientific education, and his skills were all self-taught. Charles Dickens—Gould's most famous admirer after John Ruskin and John Everett Millais—would therefore later stylize him in his journal *Household Words* into an Oliver Twist of natural history: a boy from a lowly background whose industry and character earned him the position in society he deserved.

But the romanticized tale of the rise of a poor gardener's son obscures the fact that the name "Gould" stood not just for an individual, but an entire operation. He was a magnate of the illustrated book market—the "talent scout and impresario of the feathered world," according to David Attenborough.[39] Behind Gould stood a carefully organized image factory that owed its existence to the revolutions in printing technology that occurred at the beginning of the nineteenth century. Complex techniques requiring years of training to master were replaced by other, much easier ones. Earlier, Bewick's woodcut approach had simplified the printing process but did not eliminate the need for skilled specialists. The new art of lithography, however, was easy to learn and required no professional training. Lithographs quickly replaced far more work-intensive woodcuts, steel engravings, and copperplates, especially in the large-format, illustration-dominated works that were Gould's specialty.[40]

The ease of this technique for laypersons is illustrated by the frontispiece to *The Art of Drawing on Stone*, an introduction to lithography that was popular in England. The picture shows an elegantly dressed young woman seated at a desk and engaged in the process of creating a lithograph. A finely pored piece of limestone serves as her plate. To make a lithographic plate, the artist draws on the stone's surface with chalk or ink containing grease. Because the rest of the stone is treated with a grease repellent, color applied to the plate adheres only to the lines or surfaces that form part of the drawing. The picture of the young woman in the frontispiece immediately communicates that lithography is an easy, clean craft that even women—in other words, untrained laypersons—can learn. In Gould's operation, this role was filled by Elizabeth Gould, who learned the art of lithography at the request of her husband.[41] Until her premature death in 1841, she was the most industrious cog in the Gould machine.

Several steps were involved between the design of a lithograph and the final product. Gould's contribution was to make the initial sketches that captured the animal's pose and position on the page in a few broad strokes. The sketch served as a guide for a far more detailed drawing, which was then transferred to a stone plate. Once the lithographic reproductions were made at one of London's many printing houses, Gould had the stones destroyed to ensure that the specified number of limited editions of his works was not exceeded. The lithographs were given to colorists, who meticulously daubed red on the hummingbirds' tiny breasts or azure blue in rings around the toucans' eyes. To fill the many roles needed to produce such pictures, Gould built up a staff over decades, united under his name as if it were brand. Not all the participants in the process worked directly at his offices, but they were associated so closely with him that other clients interested in engaging their skills knew they had to check with him first. To avoid losing his valued employees to competitors, Gould often kept their names and addresses secret.[42]

The division of labor Gould introduced into his book production process allowed him to produce multiple works simultaneously. The presence of his name on the printed sheet often indicated only that he had approved the image it contained. Gould let others do the drawing, create the lithographs, and color the images, but as a critic he was irreplaceable. His lead pencil unfailingly noted an incorrectly curving beak, eyes of the wrong color, proportions that were out of kilter, or a mistaken detail in the plumage. Aware that his initial sketches were inadequate, he often provided written notes about improvements the final drawings should contain. "The white stands out too far here like ears," he noted on a preliminary gouache of two dotterels. "All parts to be more neatly drawn," he wrote on the sketches of green woodpeckers before handing them off to be completed in detail. He also carefully reviewed the work of the colorists. "By no means increase the delicate citron yellow rather less if anything," he warned one employee, adding "black of bill richer" on another sheet.[43]

The fact that Gould corrected others' drawings more than he produced his own led to the rumor that he could not draw well himself—a judgment presumably based on an anecdote recorded by the son of Sir John Everett Millais. According to the story, Millais, who had become famous as one of the founding fathers of the Pre-Raphaelite movement, visited the elderly Gould in his studio at the end of the 1870s. A drawing of a hummingbird was sitting on a desk. Not wanting to admit that one of his employees was the artist, Gould pretended to be working on the piece himself. When Millais asked if the picture were finished, Gould replied: "Oh no; I am just going to put in another humming-bird in the background." However, the animal

Figure 8. John Gould in *The Ruling Passion*, an 1885 oil painting by John Everett Millais

that he was supposedly adding to the picture could not compare with the finished one. In fact, Millais's son, who had accompanied his father on the visit, reported that it did not resemble any creature ever seen on land or at sea.[44]

This incident did not reduce the admiration the elder Millais felt for Gould. Four years after Gould's death, the artist immortalized him in the painting *The Ruling Passion* (Figure 8), which was displayed at the Royal Academy in 1885. It shows the elderly Gould, surrounded by his family, lying on a chaise longue scattered with stuffed exotic birds. He holds up one small specimen and describes it to his children. His legendary bird collection—totaling more than thirty-eight thousand specimens at his death—overflows from the crates at his feet and fills the shelves and tables in the background. The scene resembles depictions of Francis of Assisi's sermons to the birds, transplanted into a museum-like settling. The tender enthusiasm with which Millais invests the frail old man caused John Ruskin, the professor of art who championed the Pre-Raphaelites, to regret that he had not become an ornithologist.[45]

The stuffed birds in Millais's painting also remind us of another aspect of ornithology: the role taxidermy played in the field. Swainson, author of both *The Natural History and Classification of Birds*, the identification handbook admired by Darwin, and *Taxidermy*, a manual of museum exhibition practices, claimed that every naturalist should be a master of both drawing and preserving specimens. Aided by his hand, eye, pencil, and notebook, the researcher should capture not only the general appearance of an animal, including its movements and proportions, but also the smallest details: "the length and disposition of the bristles with which in some tribes the bill is surrounded," for example, or "the exact shape of the scales on the foot of the bird." Swainson also considered "a quick and discriminating eye" to be essential—one "capable of perceiving at once minute distinctions." Drawing provided a way to develop such visual skill.[46] In Swainson's system, the sense of sight was akin to that of touch: in his view, a naturalist's vision is informed by what is transmitted through both the tip of his pencil and the tip of his finger. This practice is what hones the ability to perceive what makes organisms unique and elevates the craft of taxidermy into the foundation of zoological science: "It is an art . . . absolutely essential to be known to every naturalist."[47] Sensing fingers inform the eyes about the form of a preserved specimen just as the hand does as it draws. Each ability—to see, to draw, to feel—requires confirmation or correction from the others. The same interplay of senses already characterizes the study of birds in Daniel Chodowiecki's 1772 copperplate *The Ornithologist* (Figure 9). Drawings and illustrated books litter the floor of the ornithologist's study, while entire stuffed birds and other items—a peacock feather, a claw, a nest with eggs—populate the cabinet and table. The ornithologist sits with his back to the window with its fluttering curtain, a bird specimen in his right hand and book open to an illustration in his left. The specimen serves as a model for the drawing, while the drawing in turn is

Figure 9. Daniel Chodowiecki's 1772 copperplate engraving *The Ornithologist*

a guide for the specimen—a interplay of the two disciplines in which Gould also achieved mastery.

Among the subscribers to his extravagant ornithological books were numerous European royal families and university libraries, as well as esteemed scientists such as Alexander von Humboldt.[48] Gould's rivals in the ornithological book market, such as John James Audubon—whom the entrepreneurial Gould was outperforming within a few years—envied his close ties with the nobility: their love for Gould's richly ornamented world of birds guaranteed good sales for his luxury editions. But specialists who were unable to afford the books accused Gould of working for the benefit of the upper classes rather than of science. Despite this criticism of his business practices, however, the scientific value of Gould's books was never questioned.

Once Gould was made curator of the Zoological Society in 1829, his daily work provided him with an ornithological education that grew as each specimen arrived from points around the globe. His collecting fever was legendary, and Ruskin described his private collection as unmatched in Europe.[49] Countless times Gould turned an animal riddled with buckshot into a useable specimen: cleaning the throat of the dead creature with a piece of cloth, stuffing the beak and ears with cotton, slitting the body from the rib cage to the middle of the stomach and removing the organs and skeleton. His hands were skilled in loosening fat and muscle from the skin with a scalpel, unfolding the wings, picking out the bones, separating the body, and pulling off the skin over the head. This could only be performed once the eyes had been pushed from the sockets and replaced with cotton as well. Gould also knew how to form a new body from wire and flax, push the cleaned skull back up through the neck into the proper position, insert glass eyes, and use tweezers to carefully pull the semicircular eyelids down into place.[50] His hands had felt countless birds from without and within—their feathers, muscles, bones, claws, wings, and beaks. There was no part of a bird that he had not held in his hands a thousand times over. At the same time, his work with drawings—both his own sketches and the tireless corrections to the work of others—further trained his sense of sight. In Gould's case, the practice of drawing and correcting honed his vision into a model of the zoologist's indispensable tool, a "quick and discriminating eye." In choosing Gould to identify the specimens in his bird collection, Darwin drew on the hands and eyes of the man who was likely the best ornithologist in England.

Thanks to this role, Gould possessed a range of qualities that distinguished him from a ship's naturalist like Darwin. When Chodowiecki showed his bird expert with his back to a curtain-covered window, he portrayed the ornithologist as a housebound scholar—the very opposite of a researcher in the field. Gould worked indoors in the offices of the Zoological Society, not outside on expeditions; he studied dead or preserved animals, not living ones. He did not see individual animals in the wild but instead within the context of a large collection, and compared them using specimens and pictures. Darwin gathered experience in the course of his travels, but Gould developed his skills by staying at home. This very lack of mobility allowed him to see a greater number of birds than Darwin did during his entire journey. More species of birds were to be found in London in the care of the "Superintendent of the Ornithological Collection"—the position Gould reached while Darwin was sailing around the world—than anyplace else on earth.[51]

The very different paths Darwin and Gould took closely reflect the social classes to which they belonged. Born in 1809 as the fifth of six children in a prosperous, famous bourgeois family, Darwin enjoyed a privileged life from childhood. His father, Robert Waring Darwin, was a respected and successful doctor, and his grandfather Erasmus Darwin was an Enlightenment writer, scholar, and naturalist whose pre-1800 work *Zoonomia, or the Laws of Organic Life* expressed the idea of evolution in poetic form. His mother was a member of the Wedgwood family—owners of the famous porcelain business—and Darwin further cemented this relationship by marrying his cousin Emma Wedgwood in 1842.

Following his father's wishes, Darwin first studied medicine despite his sensitive nature, which was little suited to the field and the dissection of cadavers it required. He soon gave up the goal of becoming a doctor, but while still in Edinburgh he developed a keen interest in natural history—an interest he continued to pursue even after his father sent him to Cambridge to become a theology student. This time Darwin completed his studies and earned his degree.[52] When the twenty-seven-year-old Cambridge graduate, adventurer, and budding naturalist met Gould at the Zoological Society in 1837, his future was still uncertain. His professors at Cambridge would have predicted a career in geology. After abandoning the hope that his son would follow in his footsteps, Robert Darwin expected the young man to become a country pastor. This vision also seemed the most likely to Darwin himself, even during his journey. Shortly after his departure he began to build "geological castles in the air," and he decided to pursue a major geological project, even referring to himself in his letters home as a geologist. This self-image presented no obstacle to a career in the church, however. Like many a "clergyman naturalist" before him, he could have pursued his studies, led geological excursions, and gathered collections of the local flora and fauna, all the while fulfilling the duties of his office.[53] But when Darwin met Gould in early 1837, the encounter set a giant wheel into motion. The seemingly simple task of assigning the specimens in the bird collection to a species led to the further question of where species come from in the first place.

On January 4 Darwin delivered his collection to the Zoological Society. The specialists got to work on January 10. The next day, just two days after the society met officially for the first time to discuss the *Beagle* collection, the young scientist's name was in the newspaper. Mr. Darwin had presented the Zoological Society with "80 mammals, 450 birds, at least 150 different species in all," reported the *Morning Herald,* the *Standard,* the *Morning Chronicle,* and the influential weekly *Athenaeum.* Mr. Gould had already classified and named eleven species of birds, "all of which were new forms, none being previously known in this country."[54]

The eleven species mentioned in the press were all Galápagos finches (see Figure 13). In the final printed account that appeared in the *Proceedings of the Zoological Society*, Gould amended the original eleven species reported in the papers to include another three.[55] He then combined these fourteen species into a single genus, that of the "ground finches," consisting of the primary genus *Geospizinae* and the two subgenera *Camarhynchus* and *Cactornis*. The taxonomic criteria used to classify the thirty-one bird specimens Darwin had brought back were the shape of the beak, size of the body, and form of the skull. The mud-brown plumage all the birds sported, and their similar size, showed that the birds belonged to a single genus. This first identification proved the significance of the preserved birds: no longer just specimens in a collection, they became representatives of a new genus and demonstrated its morphological characteristics. Fourteen of the finches fulfilled this function particularly well and were given the status of "type specimens." Just as the archived prototype of the meter defined this unit of measure, these bird specimens define their species.[56] Even today, every ornithologist who believes he or she has found a new species on the Galápagos islands must compare the discovery with the type specimens.

During January, February, and March, Gould provided further sensational findings at a rate of one every two weeks. In a meeting on January 24 he presented six new species of predatory birds, two of which could also be found only on the Galápagos. Two new species of nightjar, a new kingfisher, and two previously unknown swallows—all, with the exception of a swallow from the Galápagos, from the South American mainland—followed on February 14. Then, on February 28, it was the turn of the Galápagos mockingbirds. Gould divided them into three species of a single genus, *Orpheus trifasciatus*, *Orpheus melanotis*, and *Orpheus parvulus*.[57]

Two factors spurred Gould, who unceasingly identified four hundred and fifty birds in just a few months, further classifying

two-thirds as new species. First, the identifications offered
Gould a further opportunity to establish himself among ornithological specialists—he became the expert who declared the
nuggets unearthed by Darwin to be gold. "I must not omit to tell
you," Gould wrote to a colleague after seeing Darwin's collection for the first time, "that Mr. Darwin's Collection of Birds . . .
are exceedingly fine; they are placed in my hands to describe;
some of the forms are very singular particularly those from the
Gallipagos [*sic*]. I have one family of ground Finches in which
there are 12 or 14 species all new." The more new discoveries
Gould could claim, the more his fame as a taxonomist grew.[58]

The second factor was Gould's skill as a craftsman. While the
similarities and differences among the birds hopelessly confused
the university-educated Darwin during his journey, Gould's practical experience allowed him to immediately assign the birds to a
genus. Thanks to his systematic knowledge of taxonomy, he could
immediately rule out particular bird genera that, on first glance,
the finch specimens resembled. By eliminating certain possibili-
ties up front, he could limit the number of families in question.[59]
But Gould's decisive advantage was his practical knowledge, developed by observing, drawing, and handling specimens in the
same way every day.[60] Darwin had incorrectly classified both the
mockingbirds and the finches during his voyage. The craftsman
Gould could see things to which Darwin was blind; only later
would Darwin surpass him as a taxonomist.

Gould was so skilled that he could identify new species even
from specimens consisting only of skin and bones. In his oft-
cited handbook *Taxidermy*, Swainson advised students to practice working with unknown and incomplete species by completely
dismantling a known type of bird, mixing the skin, feathers,
and bones into an unsorted puzzle, and then putting the bird
back together again. Darwin presented Gould with just such a
puzzle in the form of a South American ostrich. He had returned from his voyage with individual parts of the creature—
the remnants of a meal he had enjoyed in Patagonia. Realizing
too late that the dish featured an ornithological rarity, he could

rescue only those remnants of the bird that did not end up in the expedition members' stomachs.

Back in London, Gould managed to completely reconstruct the animal based on this mishmash of parts. On March 14 he delivered a scientific description of the flightless bird to a meeting of the Zoological Society, explaining its differences from the already known *Rhea americanii*. He also suggested naming the discovery *Rhea darwinii* in honor of the man who had so haplessly eaten much of it. Resurrected on paper, Darwin's meal came to be considered one of the most beautiful lithographs Elizabeth Gould prepared under her husband's guidance for *Zoology of the Voyage of the H.M.S. Beagle*.[61]

From Darwin's perspective, however, the unassuming Galápagos finches would soon be more valuable than the ostrich bearing his name. The thirty-one finch specimens, now laying lifeless and hollow with stuffed eye sockets and dull feathers in the drawers of the Zoological Society, had been identified as new species, along with the Galápagos mockingbird and dozens of other creatures brought back by the *Beagle*. But now that this work was done, Gould lost interest in the collection. "The vanity of species-mongers," in Darwin's condescending formulation, was satisfied once a species had been described and named.[62] For a naturalist and theoretician like Darwin, however, the work had just begun. The data from Gould's observations would be carried into new realms, as his specialist discovery of the "what" led to Darwin's larger question of "why?"

DARWIN'S CONVERSION

In Darwin's notebooks we can follow how, after meeting Gould, his ideas began to deviate from the canonical English view of natural history—a view that he had shared before and during his voyage.[63] The decisive break had to do with the question of the constancy of species; in other words, can species change, or are they fixed elements of the natural world that have remained the same since their creation? In Charles Lyell's *Principles of*

Geology, which Darwin read on board the *Beagle* and which had a lasting influence on him, he found all the arguments that Lyell, England's most distinguished geologist, listed in support of the belief that species do not change. Lyell argued that the Creator had shaped every animal to be as perfectly suited to its environment as a key to the lock it opens. As long as climate conditions do not change, fauna and flora remain the same as well. The mummified animals found in Egyptian excavations offered tangible proof of species constancy, since they showed that nothing had changed in thousands of years. Only a climatic shift such as the one that occurred during the age of dinosaurs could cause organisms to die out and others to arise—but these changes were foreseen in the plan of the Creation.[64] When it came to geology, in turn, Lyell represented a school of thought known as "actualism," still considered valid today. Actualism teaches that changes to the earth's surface are due to forces that can still be observed in the present, a view which rules out events such as the biblical flood as explanations for geological features. Lyell further maintained that minor factors in the earth's history could have enormous consequences, such as wind eroding the ground over a period of thousands of years. But he still saw no justification for the idea that finches, mockingbirds, or tortoises could change over time as coastlines, mountains, and oceans do. Fossilized remains found in the earth's layers indeed showed that some old species had died out, while others had emerged. Lyell, however, discounted the notion that the old species could have turned into the new ones.[65]

Darwin echoed this view himself while in Australia in 1836, during the last leg of his voyage. Describing an ant lion, the larva of a neuroptera similar to a dragonfly, he wrote, "What would the *Dis*believer say to this? Would any two workmen ever hit on so beautiful, so simple & yet so artificial a contrivance? It cannot be thought so.—The one hand has surely worked throughout the universe."[66] "The one hand" Darwin believed he could see was that of the Creator, who had equipped the Australian ant lion with the same hunting technique as that used by its

European counterpart. In both Australia and Europe, the larva digs a funnel in the sandy ground that traps passing insects when they slide into it. In 1836 Darwin had only one way to explain how ant lions identical in appearance and behavior could exist in Europe and on the other side of the globe in Australia: the Creator, whose unmistakable handwriting could be read in the presence of the same animal on two continents.

This note proves how little the Galápagos islands, which Darwin had visited in September 1835, originally impressed the young scientist. He visited just four of the archipelago's sixteen islands during the five weeks he spent there between September 15 and October 20. And while the islands had been spewed from the earth's interior millennia earlier and were never attached to the mainland, Darwin displayed little interest in their idiosyncratic flora and fauna. Since, in keeping with the theory of the constancy of species, he assumed every animal was perfectly adapted by God to its environment, he did not imagine that the species could vary from island to island. Long after his return, he continued to blur the distinction between individual islands by calling the entire group "a little world within itself."[67] Furthermore, the preliminary catalogs Darwin prepared during his voyage to keep track of the plants and animals he collected show that he recognized only six of the thirteen finch species on the island as finches in the first place: he listed one finch as a blackbird, another as a warbler, and a third as a wren.[68] It was Gould who first alerted him to these mistakes. Darwin also failed to note the locations where he had found most of the finch specimens. To his great regret, it was thus impossible for him later to specify how the finches varied from island to island. The original habitats of the individual species could no longer be identified.

Darwin's initial classification of the mockingbirds was also incorrect; in the catalog he kept aboard the *Beagle*, he entered them as three variations of a single species. Again, it was Gould who corrected this error: he declared them to be three species— *Orpheus trifasciatus, Orpheus melanotis,* and *Orpheus parvulus*— that constituted a genus. Even the remark of the archipelago's vice governor, who told Darwin during his visit that Galápagos tortoises varied from island to island, was granted little importance. More than thirty giant tortoises were loaded onto the ship to feed the crew during the next leg of the voyage, which was to Tahiti. The animals' remains were thrown into the sea, regardless of any differences that might have existed in the shapes of their shells.[69]

In April 1836 the *Beagle* reached the Indian Ocean after its visit to Australia, and Darwin began sorting the bird collection. At that point—half a year after leaving the archipelago—the budding naturalist finally realized just how puzzling the Galápagos species were. Their characteristics seemed to fluctuate between similarity and difference in an oscillation that Darwin's systematic zoological approach could barely explain. It was the mockingbirds that caused him briefly to doubt the principle of species constancy and to remember what the vice governor had told him. He wrote, "When I recollect, the fact that from the form of the body, shape of scales & general size, the Spaniards can at once pronounce, from which Island any Tortoise may have been brought. When I see these islands in sight of each other, & possessed of but a scanty stock of animals, tenanted by these birds, but slightly differing in structure & filling the same place in Nature, I must suspect they are only varieties. [. . .]; for such facts undermine the stability of Species."[70]

When Darwin wrote these lines, the *Beagle* had long left the Galápagos islands behind but had not yet reached England. In the middle of the ocean, months away from major collections and specialists, his observations were fated to remain speculation. He also did not follow this train of thought in the notes that followed. No one knows if Darwin recalled making this notation about the mockingbirds on the *Beagle* when Gould declared the birds to be three separate species just under a year later, on February 28, 1837. But the fact that, only a few weeks after meeting Gould, Darwin dismissed the theory of species constancy in his notes and argued that species could likely change suggests that he did remember. By the time he met

Gould on March 6 at the Zoological Society, where he first learned of the ornithologist's spectacular findings, his earlier thoughts on the subject must have reemerged.[71]

A sheet of paper in the Darwin Archive at Cambridge bears witness to this meeting: both sides contain apparently hastily scrawled lists of bird species in Darwin's handwriting. The word "Galápagos" has been penciled above the lists in large, flowing letters on both the front and the back.[72] All the Galápagos birds Gould had classified are listed. The catalog starts with the Galápagos hawk and goes on to name the rare barn owl, the native flamingo, the swallow, the kingfisher, and the nightjar or "goatsucker." The mockingbirds, "3 species of Orpheus," appear in the lower third of the first page. On the back at the bottom of the page we find the finches, numbered from one to thirteen.[73] Looking back on the events a year later, in 1838, Darwin would write, "In July opened first notebook on 'Transmutation of Species'— Had been greatly struck from about month of previous March on character of S. American fossils—& species on Galapagos Archipelago. These facts origin (especially latter) of all my views."[74]

Darwin specifies two groups of specimens from the collections as having awakened his interest in the possibility that species could change: the South American fossils and the Galápagos species.[75] In keeping with this remark, he also named the fossil sloths and armadillos and the Galápagos mockingbirds in the notes that surround his first sketch of the theory of evolution, in Notebook B from the summer of 1837.[76] The evolutionary diagram (see Figure 12), to which we will return, shows an original species branching into thirteen new ones—and thus contains exactly as many species as the genus of Galápagos finches does. Darwin's diagram in Notebook B contains thirteen end points, the sheet from the Zoological Society lists thirteen species, and Darwin mentions thirteen species in the published version of his travels in 1839 and in *Zoology of the Voyage of the H.M.S. Beagle* in 1841. However, since he does not actually mention the finches in Notebook B, these coincidences can do no more than suggest that he was thinking of them when he made the sketch.

More than anything else, Darwin's protracted efforts to find out which of the archipelago's islands was home to which type of finch show how important he considered them. While he had noted the locations where he had found the mockingbirds, most of the labels on the finches were missing this information—an oversight that would later prove costly. In order to learn which finch had come from which island, he turned to the collections acquired by the captain and two other members of the crew. Together with his specimens, the collections contained a total of fifty-six finches. Darwin compared the locations where they were found and furthermore send Captain FitzRoy a questionnaire.[77] But the finches were lost to him as an official example of his theory of evolution when he sailed away on that day at the equator without noting where he had found the animals. Without information on their respective environments, he could not determine why a species had split off and settled in new surroundings. As Sulloway convincingly argues, this was the reason why Darwin never returned to the finches in his works of evolutionary theory after 1859, despite their important role in his thinking. A letter he wrote in 1863 to the English explorer Osbert Salvin, who was planning a voyage to the Galápagos, shows that his wish to identify the finches' home islands persisted twenty-five years after his voyage there. Darwin wrote: "Under a purely scientific point of view, I think it would be scarcely possible to exaggerate the interest of a good collection of every species rigorously kept separate from each island."[78] But Darwin did not allude to his oversights, to the mistakes he made classifying the specimens, or to Gould's decisive contribution to the research on the birds in either the two editions of his travel book or his autobiography. As a result, he never mentions the true "discoverer" of the Galápagos finches.

THE GALÁPAGOS FINCHES BECOME ICONS

In August 1837 Darwin signed a contract to publish the zoological and geological writings resulting from the *Beagle* voyage in a series of volumes with Smith & Elder, a house specialized in large-format illustrated books. He had already sold the rights to his travel journal to the publisher Henry Colburn, of Marlborough Street. Darwin engaged the same renowned specialists who had helped him classify the specimens to work on the books, including Richard Owen for the fossils and John Gould for the birds. However, the Goulds left England on May 16, 1838, for an ornithological expedition to Australia—a continent that remained largely unknown despite Darwin's brief visit and the voyages of Cook.[79] Preoccupied with preparations for his trip, Gould devoted little time in the remaining months to the *Beagle* collection. Darwin could barely conceal his anger in his introduction to the volume on the birds in *The Zoology of the Voyage of the H.M.S. Beagle Under the Command of Captain FitzRoy:* "Owing to the hurry, consequent on his departure for Australia . . . he [Gould] was compelled to leave some part of his manuscript so far incomplete, that without the possibility of personal communication with him, I was left in doubt on some essential points."

As we have seen, one set of unanswered questions concerned the individual island to which each finch should be assigned. For example, the words "Charles Island?" mark the line indicating the habitat of *Camarhynchus crassirostris,* and even when the book indicates the locations without actual question marks, the information is not always certain. Darwin also seems unclear about the criteria Gould used to identify the various species and how he distinguished them from mere varieties.[80] As much as Darwin scolded Gould for his lack of commitment, he praised Gould's wife, Elizabeth. He had commissioned fifty lithographs from her for the ornithological volume of *Zoology of the Voyage of the H.M.S. Beagle.* These were, as the book's foreword states, "taken from sketches made by Mr. Gould himself, and executed on stone by Mrs. Gould, with that admirable success, which has attended all her works."[81] The Goulds were already on their way to Australia when Darwin received the finished pictures. They had left the completed lithographic plates with

Gould's employee Edwin Charles Prince, who was charged with overseeing their printing.

Elizabeth Gould's lithographs in *Zoology of the Voyage of the H.M.S. Beagle* captivate the viewer with their grace and precision. With the exception of four birds of prey, a goose, and the ostrich *Rhea darwinii*, all the birds are shown life-sized. Since most of the birds collected during the *Beagle*'s journey were relatively small, the quarto format used for the rest of the volumes was large enough to accommodate this. The birds were depicted in color and positioned within elements of the landscape—on a twig or a stem, in a patch of meadow, or on a cliff. The drawings came to life through details such as the upraised claw of *Milvago albogularis*, a South American bird of prey. The wind ruffles the plumage on the bird's legs as it stands on a high plateau before a background of milky blue mountains. Many birds appear two to a page to illustrate the differences between the males and females. Page after page, they teeter on branches, lean forward, or appear to whip their heads around, easily startled and poised for takeoff. One of the book's high points was the *Rhea darwinii*, with its long, airy feathers clothing its torso in gathered flounces. Three illustrations were devoted to the mockingbirds; eight of the fourteen finches also make an appearance (Figure 10).

The major deviations among the species are apparent from the lithographs alone: the beak of the compact *Geospiza strenua* is as massive as a nutcracker, while that of the delicate *Geospiza parvula* is small and pointed, and that of the subgenus *Camarhynchus psittaculus* overlaps at the front. Confronted by Elizabeth Gould's lithographs, a careful reader might think of the passage in the first edition of Darwin's travel account from 1839, in which he described the Galápagos finches and mentioned the differences in their beaks: "I have stated, that in the thirteen species of ground-finches, a nearly perfect gradation may be traced, from a beak extraordinarily thick, to one so fine, that it may be compared to that of a warbler. I very much suspect, that certain members of the series are confined to different islands; therefore, if the collection had been made on any *one* island, it

PLATE XXXVII

Geospiza strenua

Figure 10. Lithographic plate from *Zoology of the Voyage of the H.M.S. Beagle* showing the male and female finches *Geospiza strenua*

would not have presented so perfect a gradation. It is clear, that if several islands have each their peculiar species of the same genera, when these are placed together, they will have a wide range of character."[82]

Among the attentive readers of Darwin's publications on the Galápagos islands was Scottish writer Robert Chambers. When Chambers anonymously published *Vestiges of the Natural History of Creation* in 1844, he offered a theory of evolution positing that an inherent drive compelled species to develop higher forms continuously—fish becoming reptiles and reptiles birds, for example. The Galápagos islands also play a role in illustrating his system.

Vestiges of the Natural History of Creation was panned by experts but still became a best seller, producing many editions and selling more copies than *Origin of Species* as late as 1891. In the book Chambers claims that the archipelago is at a low stage of development because it can boast no native mammal species. He viewed a lack of mammals as a sign of an early stage in an evolutionary process that progresses teleologically from fish to mammal. "Latterly Mr. Darwin has discovered," Chambers wrote, "in the reptile-peopled Galapagos Islands, in the South Sea, a marine saurian from three to four feet long." According to his theory the Galápagos islands were stuck in a primitive era.[83] Despite their shared conviction that species emerge as a result of evolution, Chambers and Darwin thus differed on many points, which will be further examined in the next chapter. When it came to the Galápagos islands, Chambers's conclusion was diametrically opposed to Darwin's: the distinctions within the genus of finches showed Darwin that, far from being in an early stage of development, the island's animal populations found themselves in the midst of ongoing, species-producing evolutionary activity. Furthermore, the wealth of closely related species he found there enabled him to see evolution as a process of slight but continual change—a counterpoint to Chambers's saltation theory that assumed a leap from fish to reptile was possible within a single generation.[84] In his copy of *Vestiges*, Darwin marked the passage on the Galápagos islands with the words: "Remarks on isld not having mammals & less perfect life but really I need not to allude to such Rubbish."[85] We can only speculate as to whether he published the picture of the finches in response to Chambers's saltation theory of evolution and his characterization of the Galápagos islands. But the argument that he had carefully clothed as a mere possibility in the first edition of his travel book—using words like "if," "can," "could," and "would"—appeared the next year as a picture. Unlike language, images do not have a subjunctive mode. The picture showed the variations in the finches' beaks step by step, with a composition suggesting a sequence in time. Intentionally or not, the image seems like a response to Chambers.

The accounts ledger in the Cambridge archive contains a trace of the picture of the Galápagos finches: engraver John Lee was paid one pound, two shillings for preparing the print block in the summer of 1845. Darwin had changed publishers for the second edition of *Journal of Researches*, leaving Henry Colburn for John Murray, who would eventually bring out all his major works. Darwin revised his travel book for Murray, shortening a few scientific discussions and expanding other, more adventurous, passages, such as the encounter with the inhabitants of Tierra del Fuego. He also increased the number of illustrations from four to fourteen. Most of these pictures were taken from other published works, but Darwin personally arranged for two of them: one showing the head of the *Rhynchops nigra*, a South American waterfowl with an unusually shaped beak, and the other the four Galápagos finches. Both of these illustrations were prepared exclusively for the second edition.[86] Darwin himself commissioned the London illustrator John Lee—who, like his father, regularly worked for authors in the field of natural history—to prepare the woodcut for printing. He also paid Lee's bill for the work from his own pocket.[87]

Life-sized like their predecessors in Elizabeth Gould's lithographs for *Zoology of the Voyage of the H.M.S. Beagle,* the finches appear in the second edition in the course of the chapter about the visit to the Galápagos archipelago. Shown in profile, the four birds depicted fill half of the page (see Figure 2). Their heads, which were shown on separate pages in *Zoology of the Voyage of the H.M.S. Beagle,* have been brought together in a kind of collage. The view it offers thus corresponds to the approach Swainson taught readers in *The History and Classification of Birds* in 1836, when he arranged his subjects in terms of their physical characteristics and reduced them to silhouettes (see Figures 6 and 7). Looking at birds through the lens of such handbooks meant seeing these organisms as collections of separate characteristics and honing one's perception of the similarities and differences among them. But in Darwin's arrangement, this type of picture had new, wide-reaching ramifications: whereas Swainson would have simply grouped such characteristics together on a page, Darwin arranges the illustrations to show the levels of similarity among them. In Swainson's book the bird profiles could have been printed in a different order, but not in Darwin's. His picture reveals that the similarities and differences among species form a pattern.

At each step from Darwin's bird #1 at the upper left to #4 at the lower right, the beak becomes smaller and more pointed, and the head generally grows more delicate. Therefore, #1 and #4 differ considerably, but the intermediate transitions between each of the four profiles are far more subtle. Reorganizing tables of ornithological characteristics to make these increasing and decreasing resemblances visible is part and parcel of demonstrating their evolutionary potential. A new dimension was added to the view of the finches: while Elizabeth Gould's lithographs still showed the species constituting a new genus in terms of types that together exhibit the larger group's characteristics, in Darwin's hands they now revealed not just a type, but a phenomenon. Seen as a whole, the birds demonstrated the gradual modification of a species in the course of evolutionary history. Illustrating change in a picture required changing the form of the picture itself.

In the text Darwin also mentions that further links could be added between the four finch profiles, saying that other finches "with insensibly graduated beaks" exist beyond those shown. But from this point he left it to others to add to the picture and research the Galápagos finches. Despite his intense efforts, he was never able to assign the birds to particular islands, and this lack of underlying data precluded his mentioning the finches in his later works. In neither *Origin of Species, Variation Under Domestication,* nor *Descent of Man* does Darwin draw on the finches to prove that a species can split off in the course of the evolutionary process. All manuscript passages devoted to them were deleted from the subsequent publication.[88] Instead of the finches, English domestic pigeons illustrate the process of species formation in *Origin of Species.* But "the wonderful difference of their beaks" remains the focus, from the barb's blunt, wide beak to the runt's long, heavy one, and from the short, conical beak of the turbit to that of the short-faced tumbler, with a profile "almost like that of a finch." This comparison represents the final trace left by the Galápagos finches in *Origin of Species*.[89] Four thousand copies of Darwin's travel book were sold in its first two years of publication; in 1863 the explorer Osbert Salvin was the first to propose making an ornithological expedition to the distant Galápagos islands. In the above-mentioned letter of the same year, he wrote to Darwin asking for his support, which Darwin readily granted just a few days later, saying that he believed Salvin's voyage "would throw much light on variation (& as I believe on the origin of Species) & on geographical distribution."[90] When Salvin published his comprehensive article about the birds of the Galápagos islands in the *Transactions of the Zoological Society* in 1876, he included not only an illustration like the one Darwin designed, but also an assessment of the Galápagos islands as "classic ground."[91] Expanding Darwin's earlier model by several species, Salvin arranges nine types of finch in one row of four and another of five across the page, showing the differences in their beaks as they progress from large and heavy to small and pointed (Figure 11). With this picture Salvin saw himself completing the Galápagos research project Darwin had begun. "From this table," he wrote, describing the cool, clinical silhouettes, "it will be at once seen that the gradations from the largest to the smallest species are quite complete." He also supplied detailed information on the birds' locations and several tables precisely recording their sizes.[92] Despite Salvin's claim that his illustration perfectly captured the steps among the finch species, subsequent generations continued to refine the series, and it was a rare report on the Galápagos birds that did not contain a picture of the type developed by Darwin. Finch heads drawn in silhouette appear in the second study of the archipelago—an 1889 report by the American Robert Ridgway for the *Proceedings of the United States National Museum*—and in countless others that followed. Finally, the finches became famous around the world even among nonspecialists as "Darwin's Finches" in David Lack's 1947 book of the same name.[93]

adult. In this the crissum is nearly pure black, the feathers being very narrowly edged with white. The bill, as pointed out by Mr. Gould, has a distinct tooth-like prominence in the middle of the cutting-edge of the mandible on either side. The female is darker than is usual in other *Geospiza*.

Of *G. dentirostris* Dr. Habel says:—" Usually seen in groups of families, frequenting low bushes in search of fruit, the members uttering a cheerful chirping note. In the morning they visit the bushes growing near the shore; later in the day they retire more inland. I only met with this bird on Abingdon Island, where it predominates over the other species in numbers."—*H.*

Before passing to the next genus I will recapitulate the measurements of all the above species of *Geospiza*, except *G. nebulosa* and *G. dubia*, of which our knowledge is as yet incomplete, and also *G. dentirostris*, which seems to have distinct characters of its own.

	G. magnirostris.	*G. strenua.*	*G. fortis.*	*G. fuliginosa.*	*G. parvula.*
	inches.	inches.	inches.	inches.	inches.
Wing. . . 3·5		3·3–2·9	2·95–2·5	2·55–2·25	2·40–2·15
Tail . . . 2·0		2·1–1·8	1·85–1·5	1·65–1·4	1·50–1·3
Tarsus . . 1·05		1·0–0·9	0·90–0·8	0·85–0·7	0·72–0·63

Of all except *G. magnirostris* the largest and smallest measurements are given. From this table it will be at once seen that the gradations from the largest to the smallest species are quite complete, and that the only grounds for separating them at all rests upon the dimensions of the bill, where the steps are not quite so gradual. But the dimensions of the bills furnish but slender specific characters, as will be seen by the accompanying cut.

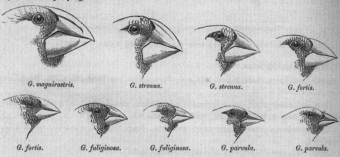

G. magnirostris.   *G. strenua.*   *G. strenua.*   *G. fortis.*

G. fortis.   *G. fuliginosa.*   *G. fuliginosa.*   *G. parvula.*   *G. parvula.*

The important and indisputable fact remains that whether we treat *Geospiza* as including one highly variable species, or as comprising several in themselves variable

Figure 11. The finches of the Galápagos islands in Salvin's illustration from the 1876 *Transactions of the Zoological Society*

The sentence from Salvin quoted above explains why this type of illustration has had a central place in studies of the birds of the Galápagos islands. It is "this table" that shows how the gradations from the largest to smallest species are practically complete. In other words: the gradation in beak shape is not a phenomenon that a traveler to the Galápagos islands could discover on his or her own, and also not one that collections of stuffed bird specimens could reveal to the unschooled eye. Before the finches could appear in the form of an illustrated table in *Journal of Researches* in 1845, they traveled vast distances and passed through many hands. Darwin and his trophy-hunting sailors shot them on the Galápagos in 1835, reduced them to skins for shipping, and conserved them in arsenic powder. In the spring of 1836, while still at sea, Darwin entered them in his preliminary catalog, mostly under the wrong names, and they sailed another half a year packed in crates to reach Falmouth and eventually travel on to Cambridge. John Gould released them from this dark prison in January 1837, unpacking them on a table at the Zoological Society, feeling and examining their tiny bodies, assigning them names and declaring them all to be finches of a single genus. Based on the specimens, Gould prepared his initial sketches. His wife developed the sketches into completed drawings and transferred the drawings to stone. Edwin Charles Prince oversaw the printing and the hand-tinting of the resulting pages. Living birds became stuffed specimens, which became species types, which became sketches, which became drawings, which finally became lithographs.[94]

The finches were turned to specimens, sent to London, examined by John Gould's hands and eyes, and captured by the tip of Elizabeth Gould's lithographic pen to emerge as two-dimensional representations on paper that Darwin cut out and arranged anew. Only after these many metamorphoses did they come to represent the process of species formation. Their visual arrangement revealed the order underlying the disorder of morphological difference.

It was the last stop in a long journey that crossed oceans, media, and specialists. The visual science of ornithology guided the destiny of the Galápagos finches in three respects: first through the activities of John Gould, whose practice of taxidermy and drawing gave him the knowledge to classify them quickly and accurately; then through Elizabeth Gould, who gave the finches their place in the visual cosmos of ornithology that had been developing between deluxe editions and identification handbooks since the beginning of the century; and finally in Darwin's rearrangement of Elizabeth Gould's illustrations—a rearrangement that led to the development of a new theory and opened a new chapter in the history of visual perception. Without the contributions of taxidermists, artists, and illustrators, the finches' gradual modification would have remained as invisible as it had been for Darwin during his voyage.

Like a telescope that enables us to see distant objects that would be invisible to the naked eye, the pictures allowed ornithology to see the gradations in the shapes of the birds' beaks. Even today, ornithological concepts preserve this link between image and research: in the twentieth century scientists began arranging the finch profiles around a center point in a circle as they developed the concept of "adaptive radiation."[95] However, this circular expansion, or "radiation"—now a term used to designate the mechanism by which one species produces several new ones—is not, in a literal sense, a characteristic of the actual birds living on the Galápagos islands. Instead, it describes how the finches are depicted in a picture. Without Darwin's picture from 1845, this genus of birds might still appear to researchers today as it initially did to the young traveler on the H.M.S. *Beagle:* "singularly unattractive in appearance."[96]

2 DARWIN'S DIAGRAMS

Images of the Discovery of Disorder

UNTIL 1859, the year in which *Origin of Species* appeared, Darwin worked in two directions simultaneously. Officially, he wrote about his experiences on the voyage, the flora and fauna in the *Beagle*'s collection, coral reefs, and barnacles—studies frequently exhibiting such an obsession for detail that the writer Edward Bulwer-Lytton caricatured him in one of his novels as Professor Long, an eccentric limpet specialist.[1] But unofficially, he was working on his theory of evolution, a comprehensive system incorporating all his observations—work that later triggered a scientific revolution with consequences well beyond traditional disciplinary boundaries. With the exception of the picture of the Galápagos finches published in 1845, Darwin long kept the public in the dark about his second project. In fact, the picture that marks the starting point of this research remained a secret for more than twenty years after he drew it in 1837.

In July of that year, after meeting John Gould at the Zoological Society for the last time, Darwin began a new page in the leather-bound Notebook B with the words "I think."[2] Notebook B was the first of his four notebooks that examined the question of species change, and it begins not with words, but with a picture (Figure 12 [Color Plate 1]). Although he verbally announces his thought process—"I think"—Darwin then switches to a picture

Figure 12. Darwin's 1837 diagram from Notebook B (Darwin Archive [Notebook B], with the permission of the Cambridge University Library Syndicate) (Color Plate 1)

to give these thoughts shape. A diagram about the size of a fist follows. From a starting point marked with the number 1, a line emerges and forks into three branches. One path simply ends, but the other two continue to split off in all directions. The proliferating lines create a fragile, irregular pattern—a sprawling growth with nodes and gaps. The letters A, B, C, and D mark the spreading sections as four groups. Those labeled B, C, and D are placed close to one another, but a large gap yawns between C and A into which one thin line protrudes and abruptly stops—a short-lived development reaching a dead end. In comments that surround the sketchlike thought bubbles, Darwin explains what the strokes, angles, and lines portray: the emergence, change, and disappearance of species. As the lines branch off, they symbolize how species vary over the course of generations. The abrupt end of a line indicates extinction. Lines that end with a cross stroke stand for recent species—in other words, those still living. "Case must be that one generation then should be as many living as now," he wrote next to the drawing at the upper left, and continued, "To do this & to have many species in the same genus (as is) *requires* extinction."[3] The number of living species thus depends on the number of extinct ones. If we count the line that trails off and those that end in a cross stroke, we arrive at twelve extinct species and thirteen living ones.

There is no way to determine whether Darwin was thinking about the thirteen species of finch on the Galápagos islands when he made the sketch. The arms of the diagram are nameless, designated in the abstract terms A, B, C, and D. What matters is the numerical relationship. The thirteen living species outnumber the twelve extinct ones by one, and this additional species is produced by competition for a constant supply of resources—what Darwin would later call the "struggle for existence." Darwin also named the mechanism indicated by the branching and disappearing lines "natural selection," adopting a principle from Thomas Robert Malthus. But even before Darwin read the British economist's *Essay on the Principle of Population* and coined the term "natural selection," he combined the elements of his theory of evolution into a picture— the stumbling, sprawling ink lines that he recorded in the summer of 1837.[4]

In all his work that followed, the diagram from Notebook B would be a motif Darwin used over and over again to depict his theory of evolution. One such diagram became the key image in *Origin of Species,* but all the others remained unpublished. They can be found today accompanying sketches, notes, and letters among his private papers in the archive in Cambridge; Darwin's last known diagram, from 1867, is reproduced in Chapter 4 (see Figure 54). In the wake of *Origin of Species,* remarked one contemporary, "family trees" shot from the earth of the natural history disciplines "with the rapidity of the fabled 'bean-stalk.'"[5] But the history that produced Darwin's diagram stretched back into the England of the first half of the nineteenth century, to the capital's zoological societies and their debate about the place of animals in the natural order. In the course of this debate, Darwin repeatedly reworked the lines, intersections, and angles of his sketches as he compared them with the visual languages of other sciences. Like seismographs, the sketches thus record the movements of his thinking as he incorporated or rejected the theories that reentered the debate in a new form after *Origin of Species* appeared. It is in the diagrams that his thoughts on the scientific theories of his day left the clearest traces; as we will see, he constructed his new picture of natural history out of models from other disciplines. In the case of the Galápagos finches, Darwin literally took four illustrations originally appearing on four separate pages and combined them into a single picture; in the diagrams, in turn, the collage was constructed in the course of the drawing itself. With a quill or lead pencil, Darwin combined elements from the highly visual tradition of natural history directly on the paper: lines indicating time, angles showing similarity among organisms, abrupt ends symbolizing extinction. As a result, the act of drawing from the past brought about something new—a surprising patchwork of tradition.[6] The debate about the system underlying nature at the beginning of the nineteenth century thus provided a starting point for Darwin to think about evolution. Knowing how to classify species was the prerequisite for asking how and why they could change.

THE STRUGGLE BETWEEN IMAGES AND LANGUAGE
TO REPRESENT THE NATURAL SYSTEM

Notebook B, which contains Darwin's diagram, takes up directly where the earlier Red Notebook left off. The contents of this earlier notebook, named for its red leather cover, date in part from the voyage of the *Beagle* and also from a brief period after Darwin's return. Back in England, Darwin filled seven additional notebooks from 1837 to 1839. He designated each of these leather-bound books with a capital letter and organized them by subject matter. In Notebook A he primarily wrote notes on geology, and Notebooks M and N contain observations comparing human and animal behavior. Notebooks B through E, in turn, are the so-called Transmutation Notebooks. Here Darwin examined the specific issue of species change—the historical process that leads to the emergence, transformation, and extinction of species.

Darwin made these notes while he lived in London and was in close contact with the city's scientific elite. The form in which he recorded his thoughts had changed from his time on the ship: back in England, a pen replaced his pencil, the pages had a portrait format rather than a landscape orientation, and the observations typically culminate in theoretical questions. On his voyage Darwin had been an explorer hurriedly capturing his impressions, holding the book by its top edge so he could write standing while propping it on his left hand. Now he was a settled scientist who wrote at a desk with ink and had the time to analyze the data collected on his travels: fifteen field notebooks, seven hundred and seventy diary pages, three hundred and sixty-eight pages of zoological observations, two hundred pages about marine invertebrates, and extensive geological notes.[7] The voyage, the rocking ship, and the foreign countries and customs gave way to a study, a desk, a library, and meetings with experts all a few miles apart in the center of London.

His new, settled life made Darwin's theoretical project possible. Until 1859, Darwin neither discussed his ideas about the theory of evolution nor confided in anyone but a few scientist friends and his wife, but he did not work in isolation.[8] While no message reached the outside world from the notebooks Darwin kept locked away, information poured in. In the collections of the city's museums, the young researcher talked with world-renowned specialists, from John Gould to Richard Owen, the

Hunterian Museum's comparative anatomist and later superintendent of the British Museum's natural history collections;[9] in scientific societies he associated with the most influential men in the field, such as geologist Charles Lyell and Cambridge professor John Henslow, who had already helped his career by recommending him as FitzRoy's companion for the *Beagle* voyage. For a time he also held the office of secretary in the Geological Society, and by 1843 he had produced not only the first edition of the account of his voyage, but also an eight-volume compendium of the zoological and geological insights it had produced. He edited the texts he had hired experts to write, gave instructions for their composition, formulated forewords to the individual volumes, added details from his notes, found artists to work on the illustrations, corrected the proofs, and made sure the project stayed on schedule.

If he intentionally avoided the topic of evolution during these busy years, it was because he knew he could hardly expect support for the project from scientific circles of the 1830s. At both the universities and the less-academic naturalist circles, the dominant view was that of "natural theology," which argued that the Creator had made all organisms, fitting them to their environments in a way both perfect and eternal. The few scientists who did not share this view could be counted on one hand. Among them were Darwin's grandfather, Erasmus Darwin, whose long, effusive poem "Zoonomia" formulated a theory of evolution that failed to catch on.[10] Two other proponents of evolution taught at the University in Edinburgh: Robert Jameson, a professor of geology and editor of the *Edinburgh New Philosophical Journal*, and Robert Edmond Grant, a physician and geologist.[11] Darwin had attended lectures and seminars by both men during his medical studies there. Also during his time at Edinburgh, the *Edinburgh New Philosophical Journal* published an anonymous review in favor of the evolutionary theory of Jean-Baptiste de Lamarck.[12] Shortly after Lamarck's death, Lyell, in his 1829 *Principles of Geology*, thoroughly discussed the ideas of the most famous champion of a theory of evolution during that time and dismissed them as not scientifically creditable.[13] As we have already seen, the young Darwin read Lyell's rebuttal during his voyage and shared his view that Lamarck's theory offered no arguments putting the constancy of species in doubt. He maintained this view throughout the entire voyage, so it seems that neither Grant nor Jameson, Erasmus Darwin nor Lamarck, made a lasting impression on him.[14]

On the contrary, it was Richard Owen and, later, John Gould in particular who convinced him, albeit unintentionally. Owen's identification of the fossils revealed that many of the extinct animals resembled species living in South America today—a correspondence suggesting that the modern species emerged from the old ones. And once Gould had classified the birds brought back from his journey, Darwin was confronted with species that differed from one another only in terms of slight deviations— the beak shapes that allowed the animals to specialize in different sources of food.

The similarity among these bird species also raised the questions of how Gould distinguished them and what criteria he used to do so. What kind and how great of a difference must exist to consider an animal as representing a new species? And when were differences simply variations among individual species members? At one point did a characteristic justify a new classification?

As the ship's naturalist on the *Beagle*, Darwin experienced firsthand how unclear the characteristics of individual animals could be and how difficult it was to classify them. But even Gould apparently did not explain to him the criteria to use when making taxonomic classifications, and Darwin's perplexity is evident whenever he deals with the Galápagos birds, whether in the third volume of *Zoology of the Voyage of the H.M.S. Beagle* or in *Journal of Researches*. Discussing the reasons for considering the various mockingbirds as species rather than variations, for example, he writes in *Zoology of the Voyage of the H.M.S. Beagle* "that if birds so different as *O. trifasciatus* and *O. parvulus,* can be considered as varieties of one species, then the experience of all the best ornithologists must be given up, and whole genera must be blended into one species."[15] However, we should remember than Darwin himself classified the birds as variations within a single species during his time on the *Beagle*. Now he maintained the opposite view, and while his evocation of "the experience of all the best ornithologists" indeed described Gould's seemingly unconscious taxonomic abilities, it was hardly a scientific justification for a particular classification.

In *Journal of Researches*, Darwin again invokes Gould whenever he refers to a species' identification: he discusses a Galápagos swallow "considered by Mr. Gould as specifically distinct," mentions a "most singular group of finches . . . which Mr. Gould has divided into four sub-groups," asks "if Mr. Gould is right in including his sub-group, Certhidea, in the main group," and so on.[16] But at no point does Darwin indicate why the birds were classified in one way and not another. Adding to the young naturalist's confusion upon his return was the fact that classifying species was not only a mysterious business to outsiders, but it had been a source of disagreement among experts for years. Hardly a reptile expert existed who seconded the work of his colleagues, no ornithologist agreed with another, and no ichthyologist was satisfied with the way his predecessor had organized the fish kingdom. The zoological community was hopelessly at odds with itself.

Darwin's frustration grew as, turning to these very specialists to classify the *Beagle* collection, he realized the extent of this discord. In October 1836, the month he returned from his journey, he was already complaining to John Henslow that he could no longer stand zoologists because of their "mean quarrelsome spirit." At a meeting of the Zoological Society one evening, he added, "The speakers were snarling at each other, in a manner anything but like that of gentlemen."[17] The arguments among the field's experts were a clear sign that a crisis was brewing, not only for Darwin and his belief in the constancy of species, but for the field of natural history in England as a whole. This crisis was the direct consequence of England's rise as a world power, and the diagram Darwin committed to paper in 1837 was his attempt to resolve the chaos that the colonial trade in animals and plants had ushered into the orderly halls of England's natural history collections.

The argument that found its way into the notebook followed the same trade and transport routes that had carried Darwin around the world on board the *Beagle*, washing up in London just like the crates full of imported animals and plants that filled its harbors. In the storage rooms of its zoological institutions, the forces that made England the world's largest colonial power took on a destructive aspect: here the growth of the British Empire was out of control. Not only were the number of Britain's colonial possessions and inhabitants and the size of its gross domestic product growing—its animal kingdom, in terms of the number of known species, was expanding to a previously unimaginable extent as well. As a result of Britain's colonial success, London possessed the largest collection of animal specimens ever assembled. The city had left its Continental rival, the Musée d'Histoire Naturelle, in Paris, far behind and had established itself as a Mecca for naturalists, attracting famous pilgrims such as Alexander von Humboldt and Christian Gottfried Ehrenberg from Berlin and Louis Agassiz and John James Audubon from America. New specimens from the colonies arrived in the capital's ports without pause. For logistical reasons, or to ensure their preservation, most of these animal remains arrived in pieces.

The masses of animal skins, pelts, skeletons, skulls, preserved organs, dried carcasses, beetles, insects, snails, shells, fish, and other items collected in His Majesty's name usually ended up in the collection of the British Museum.[18] The national museum, which had been founded in 1753, formed the hub where the often motley collections of fragments were to be converted into readily accessible knowledge. But no matter how desirable the largest possible inventory of collections might seem in theory, the reality was overwhelming. In 1825 the optimistic view asserted that the size of the collections could only help the field of natural history in England to further greatness. A glowing future was painted in vivid colors. In his euphoria, one author did not hesitate to draw a comparison with the Roman Empire: "Rome, at the period of her greatest splendour, brought savage monsters from every quarter of the world then known. . . . It would well become Britain to offer another, and a very different series of exhibitions to the population of her metropolis; namely animals brought from every part of the globe to be applied either to some useful purpose, or as scientific research." But as the animals actually did arrive from around the world, London came to resemble another aspect of the Roman Empire—the catacombs. William Swainson, the author of the ornithological handbook *The Natural History and Classification of Birds* Darwin so admired, compared the British Museum's zoological collection with the underground burial sites of Palermo, "where one is opened every day in the year, merely to deposit fresh subjects for decay, and to ascertain how the process has gone on during the previous year." Swainson had firsthand knowledge of both the catacombs and the British Museum—he had spent eight years in Italy, and he visited the national museum many times in the course of preparing his zoological publications.[19] Despite the scientific world's visions of Imperial splendor, the basement rooms of Montagu House, where the zoological collection was stored along with the British Museum's books and artifacts, came to resemble dismal graves. The collections grew too quickly for the staff to keep pace and organize and store them professionally. While exotic animals of ancient Rome were placed on show to demonstrate the empire's power over the rest of the world, England's imported animals, bloated and infested with moths, beetles, and worms, rotted away unseen and unexamined.

The daily struggle with the overflowing collections sowed overwork and frustration among the museum's curators, since they faced the greatest challenge from the flood of specimens washed up in the colonial collecting frenzy and forced down the museum's throat. It was up to them to find a place and a name

Figure 13. A Galápagos finch with a museum label filled out by Darwin (With the permission of the Natural History Museum, London) (Color Plate 2)

for each item. In each individual case, a museum zoologist had to check whether the specimen in question had already been assigned to a species and thus had a name, or if it was a not-yet-discovered species and needed one. At the same time, the animal kingdom ballooned to unmanageable dimensions. In 1825 a contributor to a zoological journal complained that the number of known species had increased fivefold since Linnaeus's day; yet just twelve years later, it had grown a hundred times over. In other words, for every animal that Linnaeus had described in his *Systema Naturae* in 1768, there were suddenly a hundred new animals, all needing identities.[20]

Naturalists had two techniques for regaining control. The first was to assign a specimen a name and enter it in the catalog (Figure 13 [Color Plate 2]); the other was to identify its position within the system of nature (Figure 14). Both techniques belong to taxonomy, the science of identifying species, but each employs a different medium. The first approach involves using linguistic labels to impose order, while the second—assigning the species a place in the natural system—literally required the taxonomist to draw a picture. The position of a particular species was depicted in complex geometric figures that characterized both the specific point in question and the system as a whole.

A broad range of such systematic pictures and diagrams existed in the nineteenth century. They took advantage of every geometric construct available, from circles, ladders, and squares to lines, stars, and brackets. As the number of nameless

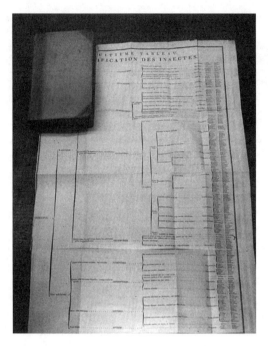

Figure 14. The fold-out table "Classification of the Insects" from the first volume of Georges Cuvier's *Leçons d'anatomie comparée,* from 1799

representatives of the animal kingdom reaching England grew by the day, the sheer variety of possible pictures and labels sowed confusion within these ordering systems. The overabundance of research material made the questions of what makes a species and how multiple species form a system the key problem confronting the country's naturalists. As they searched for an answer, the line dividing descriptive natural history on the one hand and explanatory physical sciences on the other began to blur. The need not only to describe the organic world but to understand the laws governing it now came to the fore. Scientists sought to ground the order behind the animal kingdom in mechanistic principles, based on the general laws of motion, that acted to produce life in its many variations. If God was the force behind the variety of species, then it seemed likely that He followed a plan and laws during the creation. The task of the naturalist was to re-create these laws.

English researchers were not alone in seeking the laws underlying the natural world. They were joining a scientific current that had already been set in motion on the Continent. It was no accident that France had long taken the lead in this quest—thanks to colonialism and global trade, the French had also amassed enormous collections of natural specimens. As early as 1825, Baron Georges de Cuvier, head of the Musée d'Histoire Naturelle and member of the Académie Française, prophesied that natural history would soon have "its own Newton." Coming from Cuvier, this promise had weight, since he had already amazed his contemporaries with his ability to reconstruct an entire animal—an extinct one, at that—from a single bone: the carnivorous giant lizard that was later given the name *megalosaurus*. To accomplish this feat, Cuvier rigorously followed his theory of "functional morphology," which held that the form and function of every animal was perfectly adapted to the requirements of its environment. As a result, he believed, it was possible to read any animal's identity and classification from any of its body parts. According to Cuvier, a study of comparative anatomy could yield the laws and mechanisms that determine how species develop in the animal kingdom. The field therefore went hand in hand with taxonomy, since it provided the criteria for organizing the wealth of organisms into groups as a prerequisite to discovering the laws behind their similarities and differences.[21]

In Germany, the school of so-called physicoteleologists likewise emerged with the goal of researching the mechanics of species formation. Following the teachings of the philosopher Immanuel Kant, they recognized teleology as an a priori condition for living things, believing that every development—that of an organism in its embryonic stage, for example—progresses toward a set goal. At the heart of physicoteleologic inquiry, however, was not the idea of teleology itself, but rather the search for the mechanisms that govern the development of species and organisms from one stage to the next. According to the physicoteleologic view, a naturalist seeking to discover the teleologic laws governing the animal kingdom needs understand the reasons for this teleology no more than a physicist needs to understand why gravity exists before describing the laws governing falling objects. The physicoteleologists sought the criteria for organizing the animal kingdom not by comparing the anatomy of adult animals but by studying their embryonic development. This distinction would prove decisive for the debate in England.[22]

There, mathematician and astronomer Sir John Herschel, in his work *Preliminary Discourse on the Study of Natural Philosophy*, declared that taxonomy and classification should be the primary tasks of natural philosophy in the future. The Cambridge Kantian William Whewell repeated this sentiment in *History of the Inductive Sciences* in 1837.[23] Cuvier's prophecy that a Newton of natural history was coming found an echo in countless works that likened the future of zoology with the successful history of astronomy. Like the heavens before it, the animal kingdom would be mapped, allowing the laws of species formation— like the movements of the planets—to be understood. Taxonomists thus saw themselves as the "Keplers of natural history," whose efforts to discover and describe the system of nature would prepare the way for the Newton of their field. Darwin's countrymen nursed the hope that, despite France's lead, this naturalist Newton would be English, like his namesake. Mathematician Charles Babbage, another Cambridge scholar, further fanned this longing for glory when he claimed that his country was lagging behind the transcontinental competition in the area of the physical sciences.[24] The scepter that had passed from Newton's native country to France in physics would be reclaimed from the land of Cuvier and Buffon in natural history.

As the seemingly simple question of how to identify a species advanced to an issue of national honor, the search for the order of things sparked a fierce debate about images.[25] One of

Darwin's notes from 1841 reports on the situation. On a loose sheet of paper, preserved in the archive in Cambridge, he wrote that "disputes about affinity, linear, circular arrangement and definition of species" had become a "vexed question."[26] The lines and circles he mentions refer to the countless taxonomic diagrams shown on blackboards, as posters, or in sketches at debates and later published in books and journals. Darwin himself would eventually take sides in the contest between the linguistic act of naming and the visual practice of placing animals within the natural system—a contest that, between 1830 and 1840, escalated into a debate for naturalists reminiscent of the art world's *paragone*.[27] At issue in this contest of word and image were the truth of words and the truth of pictures. How accurate were the names mankind bestowed upon the species? How accurate were the pictures mankind made of the system of nature? What criteria permit us to name an animal? Or make a picture of nature? Which medium can show the true order of nature, without distorting it—words or pictures?

When it came to names, the experts encountered some obstacles. At first the difficulty was a tangible one—the limits Linnaeus had set on the creativity of specialists working to assign names to the imported specimens. The Linnaean system, which did not allow for subspecies, was groaning under the weight of the animal kingdom's diversity. Linnaeus's approach to classification was increasingly unable to cope with the fine distinctions among species, subspecies, and varieties that became more and more apparent with every shipment that reached London. Eventually its foundations collapsed under the weight of the shiploads arriving daily. It was clear that the Linnaean catalog of criteria needed to be expanded—or completely replaced.

Suggestions for better dividing the animal kingdom and designing a new classification system came from all sides, and agreement seemed nowhere in sight. The question of whether an animal should be classified based on randomly selected external features, as Linnaeus had suggested, or on anatomical criteria such as the most important organs, as in Cuvier's method, was hotly debated. Malicious tongues spread the rumor that the president of London's respected Linnean Society, founded in 1788, had suggested solving the problem by crushing under his heel all the shells that did not appear in the tenth edition of the *Systema Naturae* or, barring that, smashing them with a hammer. A counterproposal that scientists should adopt the taxonomic principles of Jean-Baptiste de Lamarck instead of those of Linnaeus was rejected with the claim that "the trash vomited forth by Lamarck and his disciples" was nothing less than "atheism and blasphemy."[28] And while in London's ports the ships kept arriving with new plunder from the colonies, the ordering system became a web of suggested reforms, accusations, and insinuations. As various specialists uncovered mistakes by Linnaeus, the impression grew that the system was not equipped to meet the requirements of the new age. For example, the Swedish scientist had mistakenly named the omnivorous common rook *Corvus frugilegus*, or "seed-gathering" rook, and dubbed a bird *Lagopus britannicus*—British snow grouse—although it existed throughout Europe.[29] In Linnaeus's system, the name of an animal often reflected the criteria used to assign its position. As a result, classification mistakes could be eliminated only by changing the names themselves. But changing an animal's name was fraught with peril. Even assuming agreement could be reached on a new system, what would happen to the old names that would no longer be in line with it? If they were replaced with new names, previous works of natural history could end up unread. All earlier books would suddenly be obsolete— and centuries of zoological research along with them. Replacing the old names with new ones, many voices warned, would sow confusion of Babel-like proportions. The chaos of new names and old ones—correct and erroneous—would mean experts would no longer understand one another. The scientific community would splinter from generation to generation under the array of contradictory designations.[30]

After almost two decades, the debate finally reached a resolution in 1842 thanks to Hugh Strickland. The Oxford graduate and zoologist, who later wrote the much-admired monograph *The Dodo and Its Kindred* on the extinct bird of Mauritius, organized a committee for zoological nomenclature for the annual meeting of the BAAS, the British Association for the Advancement of Science.[31] Among the committee's members was Charles Darwin. Strickland's proposed solution was a pragmatic one: all names already commonly used for species should be maintained. Experts would have the authority to establish new names for newly discovered species. Whoever described an animal first and published this description in a scientific journal would determine the animal's name forever. This "law of priority" also functioned retroactively. If proof of an earlier identification of a species existed, even a commonly accepted name would be changed. The first name published was the one that counted. This rule explains why the South American ostrich brought back by Darwin is no longer known as *Rhea darwinii*—the name Gould gave the bird in 1837—but rather *Pterocnemia pennata pennata;* it turned out that a specimen of the species had already been described by the French scientist Alcide Dessalines d'Orbigny in 1834. For animals that had not yet been identified, the experts naming them could follow the system of Linnaeus, Cuvier, Lamarck, or anyone else. What mattered in the end was that the animal had a name suitable for a label in a museum. Because, according to Strickland, not the name itself but its general acceptance was what mattered.[32]

Consistency of usage guaranteed that a zoologist in London could communicate with his colleagues in Berlin, at Harvard, or in Paris. Names, after all, did not just appear on the labels attached to the specimens—they literally served as labels themselves. A generally accepted name ensured that a naturalist could look up a species in the index of a book or a museum catalog and find it. In the "Mecca of natural historians" (the British Museum), names guided the specialists to the room, cabinet, and drawer in which they could find an example of a species they sought. For Strickland, the idea that a name could be "wrong" rested on a misconception. He saw names as purely conventional—as arbitrary signs for the objects in nature. In this regard, names for species were no different from personal names. As he countered to his critics, "We do not object to William Whitehead's name because his hair may happen to be red." In the 1840s his "Rules of Zoological Nomenclature" achieved acceptance in the English-speaking world, and during the course of the century they came to form the foundation of an international obligatory system. Since that time the first name given an animal is considered valid, although identifying and naming a new species requires scientific consensus. A newly discovered species "counts" when its description appears in a scientific journal.[33]

An elegant solution had thus been reached for the problem of names. But while the question of how words and things are related may have been resolved, the debate on the order among these things themselves was far from over. Words and names could be random signs or labels for a drawer, independent of the essence or organizing system of the things to which they refer. This did not, however, mean the reverse was true—that no natural order exists among nature's objects. But scientists no longer sought to discover this natural order using words: instead, the medium they hoped would reveal the ordering principle among living things was now pictures. The flip side of the debate on names now took place in the context of images.

Strickland also played a role in refereeing the paragone's outcome as a decision against language and for pictures. Two years before the Committee for Zoological Nomenclature was appointed, Strickland presented his paper "On the True Method of Discovering the Natural System in Zoology and Botany" at the Zoology Section of the British Association for the Advancement in Science during the annual meeting in Glasgow. The "true method" he recommended was to represent the natural system as a picture. Strickland portrays this task as arduous and challenging, one only a group of researchers working collectively could hope to master. He also compiled a number of insights that could be helpful in the search for the real system underlying nature.

"It appears," Strickland began, "that the natural system is an accumulation of facts which are to be arrived at only by a slow inductive process, similar to that by which a country is geographically surveyed."[34] What followed sounds like a call to create natural history's still unknown masterpiece—as if Strickland were issuing an invitation to his colleagues to join a race. The overall shape of this great work, he noted, was still unclear, but many of its features were already certain. First, the natural system would be irregular; second, it would—in keeping with its character—consist of a visual image; and third, it would have many branches:

> Organic life exhibits . . . irregularity—no two planets, and no two leaves of the same plant were ever perfectly identical in size, shape, colour, and position. In the "human face divine," portrait-painters affirm that the two sides never correspond; . . . In short, variety is a great and a most beautiful law of Nature; it is that which distinguishes her productions from those of art. . . .
>
> The true order of affinities can only be exhibited (if at all) by a pictorial representation on a *surface,* and the time may come when our works on natural history may all be illustrated by a series of maps on the plan of those rude sketches which are here exhibited. . . . All that we can say at present is, that ramifications of affinities exist.[35]

The map Strickland drew for the conference in Glasgow is still preserved among his papers at the archive of the Cambridge University Museum of Zoology. Pasted together out of three dozen individual sheets of paper, it shows the divisions among birds in several colors and on a monumental scale (Figure 15 [Color Plate 3]). As the first chapter of this book describes, birds were the most popular and best-researched group of animals in the nineteenth century, and the large number of ornithological specialists meant that the system of connections among them was an issue of particular dispute.

Figure 15. Strickland's 1840 wall map for classifying the birds (With the permission of the Zoological Museum, Cambridge) (Color Plate 3)

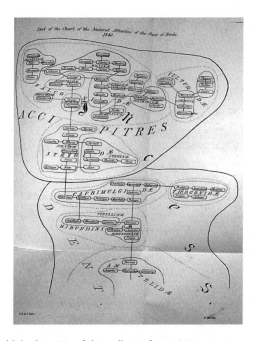

Figure 16. Published version of the wall map from 1841

At first glance, Strickland's organization of the birds makes a less-than-orderly impression, in both his original chart and the version published later (Figure 16). He distributed the various bird species over the entire surface of the drawing, inscribing their names within small ovals. He indicated kinship among species using connecting lines, whose length varied depending on the proximity between various species. Groups of species constituting a genus were encircled with thicker lines, while the genera belonging to a single family were colored yellow, green, red, or orange or, in the printed version, marked with dotted boundary lines. Strickland indicated the orders—such as that of the aquatic birds, *Natatores*—with a broad blue line that sweeps in irregular arcs around the families, genera, and species. Strickland's picture shows a natural system devoid of regularity: at

some points the species clump together in dense nodes of affinity, but in other places the species groupings become looser and
more white space appears between them.[36]

By incorporating branches, circles, and lines, Strickland
echoed the repertoire of shapes used by his predecessors in the
field of natural history, only to turn around and reject this inheritance in a second step. In itself, the idea of drawing the natural
system as a picture was not a groundbreaking one; despite the
visionary tone in which he prophesied a time "when our works
on natural history may all be illustrated by a series of maps,"
this very practice was already common. Baron Georges de
Cuvier's *Leçons d'anatomie comparée* (1799–1805), still a standard zoological work forty years after it first appeared, contained nine fold-out tables, each of which charted a division of
the animal kingdom (see Figure 14). The sheets were affixed inside the cover and could be folded out several times—in the case
of the insects, Cuvier's special area of expertise, four times horizontally and three times vertically. With just a few motions, the
table thus grew to ten times the size of the book itself, making it
nearly the size of Strickland's chart.

In a smaller format but with no less care, William Swainson
published his version of the ordering principle among birds in
1836 in his three-volume work *The Natural History and Classification of Birds* (Figure 17). Swainson confided to his readers that
he had worked to decipher the natural system of birds for thirteen years. Despite financial worries and poor health, he explains,
he visited museums throughout Europe, purchased collections
of specimens, and wrote to collectors around the world to obtain
the necessary research material. Like Strickland five years later,
Swainson protested that his system was not finalized, but he—
again like Strickland—believed he was on the right path. Bor-
rowing the stirring words of another zoologist, he wrote: "When
we cannot represent Nature as she is, we must endeavour to represent her as she appears to be; for if we suspend our observation in apprehension of committing an error, we shall soon
cease to represent her at all."[37] He then went on the describe

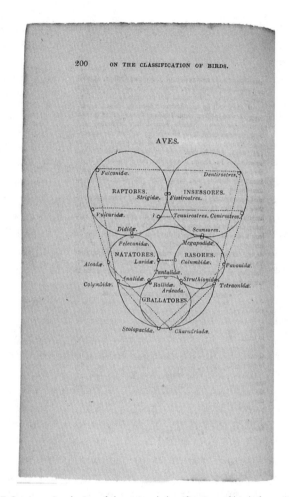

Figure 17. Swainson's scheme of the natural classification of birds from 1836 (University Library Cambridge with the permission of the Cambridge University Library Syndicate)

the fruit of his effort: his arrangement of the birds consisted of five circles stacked upon one another to form an inverted pyramid.

Despite their obvious differences, Cuvier's and Swainson's systems resembled each other in one respect: their regularity. According to Cuvier, branchings in the animal kingdom always resulted in groups of two—a binary system that held sway from the level of the order all the way down to that of the species. Following this principle, animals were classified based on the presence or absence of characteristics; insects, for example, were grouped into those with and without wings. Swainson, an opponent of Cuvier's binary approach, organized the animal kingdom not according to oppositions, but rather into circular groups of five similar animal species. Methodically, he thus followed the teachings of the "quinarians," a taxonomic movement that flourished in the 1830s, especially in England. The founder of the quinarian school of thought was William Sharp MacLeay, a London beetle expert who firmly believed that all natural groups were based on the number five.[38] The circle symbolized the imagined ringlike organization of the animal kingdom, signifying not only that species form groups, but also that all animal families are related to one another. MacLeay taught that every species, every genus, and every family passed through the same five stages of development in parallel—a belief that led Swainson, for example, to refer to the birds as the "butterflies among the vertebrates." The entire system rested on proportional mathematical relationships. "No rational being," Swainson claimed, "can suppose that the great Architect of the world has created its inhabitants without a plan." In the absence of a plan, each animal would be merely an imperfect ornament, like a column or frieze developed by an architect who fails to consider the role it would play in a building.

At another point, Swainson took up the comparison of the natural world to the heavens that Cuvier had drawn in France and Herschel and Whewell in England as they looked forward to the coming of the "Newton of the grass blade." Doubting that plant and animal forms develop in circular patterns, he declared, would be like doubting that the planets move in circular orbits.[39] His teacher MacLeay also expressed this creed of harmony, regularity, and mathematically proportional relationships as he discussed his sketch of the system underlying the *Annulosa*, a subclass of invertebrates (Figure 18). "Nothing in Natural History is, perhaps more curious, than that these analogies should be represented by a figure so strictly geometrical," he wrote. "One is almost tempted to believe that the science of the variation of animal structures may, in the end, come within the province of the mathematician."[40] While Strickland compared naturalists to cartographers, MacLeay, Swainson, and Cuvier invoked architects, astronomers, and mathematicians. Their models for future inquiries into the living world were drawn from the physical sciences.

This conviction that nature was based on a mathematical composition resulted in not only a highly geometrical visual language overall, but some very idiosyncratic formal elements. If his classification did not produce a grouping of five, for example, MacLeay simply drew three asterisks as placeholders in the empty spot, confident that explorers would soon discover animals to fill these gaps (see Figure 18). The quinarians checked printed versions of such diagrams with irritable meticulousness, as their aggravated postscripts pointing to mistakes made by their lithographers or engravers attest.[41] To them, an incorrect line, a misplaced circle, or a missing asterisk spelled the difference between truth and error. Twenty years later, Alfred Russel Wallace would recommend avoiding any symbol or form not contained in the engraver's repertoire when preparing diagrams, precisely to prevent such corruptions from occurring. The difficulties involved in printing them means that many diagrams shown as posters, blackboard drawings, or sketches at zoological society meetings were never published and were eventually lost.[42] Strickland's map at the Cambridge University Museum of Zoology is one of the few exceptions.

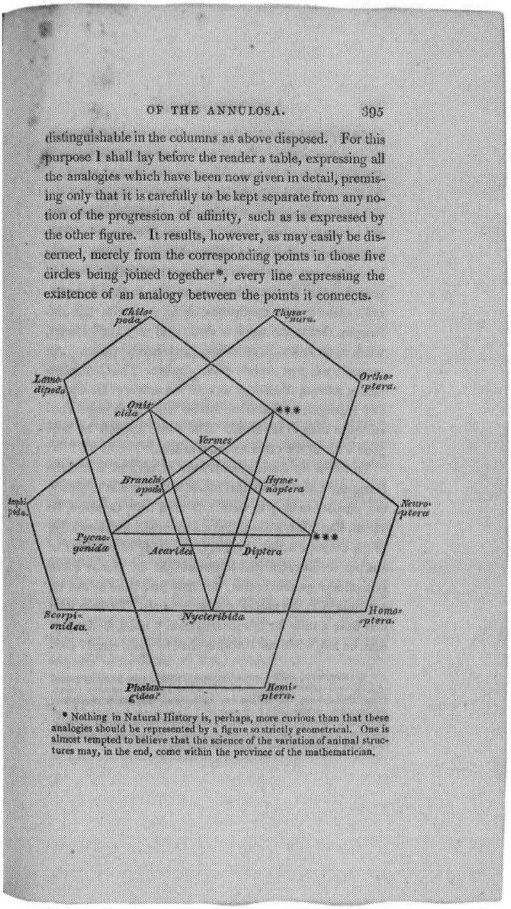

distinguishable in the columns as above disposed. For this purpose I shall lay before the reader a table, expressing all the analogies which have been now given in detail, premising only that it is carefully to be kept separate from any notion of the progression of affinity, such as is expressed by the other figure. It results, however, as may easily be discerned, merely from the corresponding points in those five circles being joined together*, every line expressing the existence of an analogy between the points it connects.

* Nothing in Natural History is, perhaps, more curious than that these analogies should be represented by a figure so strictly geometrical. One is almost tempted to believe that the science of the variation of animal structures may, in the end, come within the province of the mathematician.

Figure 18. MacLeay's classificatory diagram of the invertebrates from 1821 (University Library Cambridge with the permission of the Cambridge University Library Syndicate)

"On the True Method of Discovering the Natural System in Zoology and Botany," Strickland's 1841 paper, was therefore not new in insisting that the natural world should be presented visually. All parties to the debate gave the image precedence over the word; MacLeay even went so far as to disparage the act of naming species. He claimed that any fisherman or botanical hobbyist could assign names to animals and distinguish among them, but scientists were the only ones who could show the true organization underlying the animal kingdom and its affinities and analogies. What was new, however, was Strickland's emphasis on irregularity. His chart is a type of collage combining the old circular, crystal-shaped, and branching diagrams, but rather than presenting the symmetry of the animal kingdom it reveals disorder.

Key to his map and the related metaphor of the expedition—one in which the animal kingdom, like a new territory, would be discovered bit by bit—was the unpredictability of the enterprise. The resulting picture could turn out to be as jagged and frayed as the coast of Tierra del Fuego, which Darwin himself had once helped to map. Until now, such a lack of order had been unthinkable in taxonomy.[43] According to Strickland, there were no gaps that had to be filled; defying the irregularity of the natural system, as MacLeay had done by inserting stars as placeholders, would be like a cartographer drawing the coast of Tierra del Fuego as an algorithmic curve. Acknowledging nature's disorder became Strickland's manifesto. "When, therefore, we find a system of classification proposed as the natural one which . . . fetters the organic creation down to one unalterable geometrical figure or arithmetical number, there is, I think, a strong a priori presumption that such a system is the work not of nature but of art."

Regularity was thus artificial, counterfeit nature, while its absence was natural; with this claim Strickland overturned the previously held parameters of natural history. Until now, order and harmony had been considered the fundamental features of nature, and the more a researcher discovered organized structures, the closer he seemed to the truth. But Strickland declared that the presence of order automatically meant a system could not be valid. Instead of geometric figures, the natural system he represented in 1841 took the form of a map, and organic metaphors describe it at the level of language: "The natural system may, perhaps, be most truly compared to an irregularly branching tree, or rather to an assemblage of detached trees and shrubs of various sizes and modes of growth."[44]

In the silence of his study, the young naturalist Charles Darwin had also taken part in the search for the image of the natural system for several years, making small ink drawings in his notebook. Back in 1837 he had sketched branching diagrams in Notebook B and commented: "Organized beings represent a tree, *irregularly branched,* some branches far more branched."[45] His scrawled little drawings captured the impressions he had gathered in London's museum collections: the disorder and lack of intelligibility, the uncontrolled growth that had turned the storage rooms of zoological collections into catacombs. Looking back on his work with the zoologist charged with identifying the items in the *Beagle's* collections, he wrote in *Origin of Species:* "Many years ago, when comparing, and seeing others compare, the birds from the separate islands of the Galapagos Archipelago, both one with another, and with those from the American mainland, I was much struck how entirely vague and arbitrary is the distinction between species and varieties."[46]

During the summer of 1837, amid London's zoological collections, Darwin began to doubt that species are unchanging units within the natural system. He saw evidence of their mutability in the day-to-day business of taxonomy, as species and varieties seemed to change each time a new expert examined them. The thirteen finch species identified by Gould could be grouped by the next specialist into just twelve, and zoologists today consider them to represent just nine species.[47] And the boundary between species and variety became increasingly blurred as more and more animals arrived in London.

A rift opened up in the zoological world. Some scientists believed that new, better criteria must be found to allow clear distinctions to be made and restore order; others, including Darwin
and Strickland, began to trust the existing chaotic state of the
zoological collections more than the promise that order could
be reestablished at some future point. Rather than distorting
human perception of nature, this very lack of order seemed to
represent it. But despite their similar views, Darwin did not tell
Strickland about his efforts to develop a theory of evolution. He
failed to mention it during their short but intense correspondence in 1842 about scientific nomenclature, which Darwin believed should be standardized as Strickland proposed; likewise,
he maintained his silence on the subject when he began to classify the barnacle family in the late 1840s and often turned to
Strickland for advice. The only evidence that Darwin was aware
of their similar views comes from a will he made in July 1844, in
which Darwin named Hugh Edwin Strickland as one of the ad-
ministrators of his scientific papers in the event of his unexpected death.[48]

EMBRYOLOGICAL DIAGRAMS

On his journey toward the discovery of a new way to imagine
nature, Darwin—like many of his contemporaries—initially
found shelter in one of the most popular classification systems
of his day. His notebooks refer more than two dozen times to
MacLeay, whose approach appealed to Darwin because it organized the natural world based on the similarities between animals.[49] Unlike Cuvier's system, for example (see Figure 14), which
sorted families, genera, and species according to the differences
among them, MacLeay's geometric scheme is based in a system
of interlocking affinities and analogies: so-called osculant—
literally "kissing"—groups formed the transitions between separate groups of animals. The metaphor of the kiss referred to
the placement of some of the circles in his diagrams: wherever

two circles touched, they indicated an analogy between the categories that constitute the animal kingdom. MacLeay, for his part, had drawn on the work of William Smellie. In his 1799 work *Philosophy of Natural History*—the same book in which Darwin marked the passage about the "circle of animation and destruction"—Smellie also devoted a chapter to the "progressive scale of chain of beings in the universe." Darwin would translate this ideal ladder into historical terms, breaking its stages into the countless steps of a horizontally expanding labyrinth.[50] Before it led to the diagram from Notebook B referred to at the beginning of this chapter, Darwin's private debate with MacLeay produced an initial, more modest attempt at visualization: a simple sketch barely four centimeters high (Figure 19 [Color Plate 4a]). It shows a dotted line that branches into three solid lines at the top. The sketch represents the three-part system of nature described by MacLeay in which the animal species are to be divided according to the elements: "A triple branching in the tree of life owing to three elements air, land & water," as Darwin had noted three pages earlier.[51]

When Darwin transferred MacLeay's vision of an animal kingdom divided according to the three natural elements into an image of a tree with three branches, he introduced the dimension of time into the system. The preceding notes almost exclusively concern extinct species, including the giant sloth *megatherium,* whose close kinship with living sloths made Darwin aware of the similarities between extinct and nonextinct species in South America. "We may look at Megatheria, armadillos & sloths as all offsprings of some still older type," he wrote. In explaining this resemblance as the result of a common origin, Darwin was forced to correct MacLeay's theory. "The tree of life should perhaps be called coral of life, base of branches dead; so that passages cannot be seen."[52] For this reason, Darwin drew the trunk as a dotted line, introducing a convention from geology into his classificatory diagram. Geographers used dotted lines on maps to indicate the location of coral reefs that formed the foundation of an atoll under the water's surface

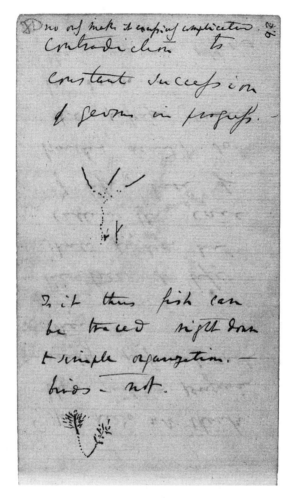

Figure 19. Two diagrams in Notebook B occurring a few pages before the "I think" sketch (Darwin Archive [Notebook B], with the permission of the Cambridge University Library Syndicate) (Color Plate 4a)

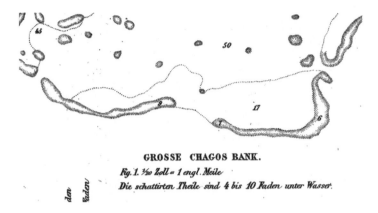

GROSSE CHAGOS BANK.

Fig. 1. ⅟₅₀ Zoll = 1 engl. Meile

Die schattirten Theile sind 4 bis 10 Faden unter Wasser.

Figure 20. Excerpt from one of the maps in Darwin's 1842 *The Structure and Distribution of Coral Reefs*

while remaining invisible to the human eye.[53] In 1842 Darwin had drawn these submerged features as broken lines on the map in *The Structure and Distribution of Coral Reefs*, and now he used the same sign to indicate the sunken remains of the animal kingdom hidden beneath the earth's crust (Figure 20).

The shape of the tree of life thus is not that of a single coral, but rather the interior structure of an entire coral reef. Next to the sketch Darwin depicted the smallest unit of evolutionary system: a short forking line. In the drawing that followed he used the same image to apply this just-discovered idea to two different groups of animals (see bottom of Figure 19), writing, "It is thus fish can be traced right down to simple organization.—birds—not." The one arm of the forking line crowned by feathery ink lines is thus solid, while the other is dotted once again. The solid line stands for the many fossil fish that already documented the history of this class of animals in the 1830s. The broken line, in turn, shows the fossil record for the birds, which was full of gaps. Their history remained invisible to the researcher and thus hypothetical.[54]

At this point, a leap is evident in the notebook. In the famous picture that follows—the diagram with the heading "I think"—

the basic evolutionary unit which Darwin, in passing, had jotted down ten pages earlier now appeared over and over. Variation and extinction turned the earlier orderly little tree into a mass of tentacles, fragile and difficult to follow. Instead of three branches, there were now four, and capital letters divided the splintering image into four groups.

Darwin's switch to a shape based on four elements rather than three offers clear evidence of the exchanges he had with scientific experts during his London years, especially with Owen, the anatomist at the Hunterian Museum who identified the fossils collected during the *Beagle*'s voyage. Owen was likely the one who made Darwin aware of a sensational new theory about the variety of organic forms: the four archetypes proposed by the German embryologist Karl Ernst von Baer, which the Scottish physician Martin Barry introduced to English-speaking readers in a two-part essay in the *Edinburgh New Philosophical Journal*.[55]

The installments of Barry's essay appeared in January and April 1837, shortly before Darwin began his notebook on the question of species change, and their appearance introduced German idealist morphology into English science. In the course of the debate about the natural system, Barry drew upon Baer to suggest classifying the animal world according to attributes seen in embryonic development. He claimed that the underlying principle by which the organisms were organized was not adaptation—that is, that the animals had been designed by the Creator to function in their environment—but rather form. In this view, the animal kingdom could be divided into ideal forms or archetypes that were also evident in the stages of embryonic development. According to Barry, Baer's embryological studies had proven "that in all classes of animals, from Infusoria to Man, germs at their origin are *essentially the same in character;* and that they have in common a homogenous or general structure."[56] This means that every animal starts out the same and acquires typical taxonomic characteristics as it develops. Embryology was thus key to the classification of organisms. Not

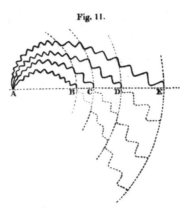

Fig. 11.

Figure 21. Barry's diagram from the 1837 *On the Unity of Structure in the Animal Kingdom*

as an adult, but only in its developing form, could an organism provide information about the similarities among animals.

Barry explained these findings using a number of unusual diagrams that presented both this common origin and the gradual differentiation that followed in visual terms. His first diagram demonstrated origination from an archetypal form using several wavy lines that emerge from a point labeled A and then radiate outward (Figure 21). The letters B, C, D, and E stand for the four types of vertebrates: fish, reptiles, birds, and mammals. The points where the waves intersect the horizontal axis indicate the stage at which the embryos most resemble one another. With each day of further development, the organisms grow more and more different and the waves spread further and further apart. The second diagram shows the increasing differentiation among classes of animals in the course of the organisms' development (Figure 22). Following a tradition of natural history, Barry called this diagram "The Tree of Animal Development," but the image resembles a tree only insofar as it contains branching structures. A trunk, for example, is nowhere to be seen in the diagram. The passage from general to *ner* of the change, is probably the same throughout the animal kingdom, however much, 3*dly,* The *direction* (or *type*) and *degree* of development may differ, and thus produce variety in structure; which, however, there is good reason to believe, is, 4*thly,* In essential character, *fundamentally the same.* Yet, 5*thly,* That no two individuals can have *precisely* the same innate susceptibilities of structure, or plastic properties; and therefore, 6*thly,* That though all the individuals of a species, may take, in their development, the same *general* direction, there is a *particular* direction in development,—and, therefore, a *particular* structure,—proper to each individual. 7*thly,* That structures common to a whole class must, *in a modified form,* re-appear in individual development; and, *lastly,* That they can re-appear in a *certain order only*; viz. in the order of their generality in the animal kingdom.

These conclusions, especially the two last, with the reasoning from which they are derived, sufficiently explain why, in the embryonal life of the more elaborate animals, there occur temporary resemblances in certain parts of structure, to the permanent states of corresponding parts, in animals less wrought out.

A Diagram will serve to illustrate some of these conclusions.

The Tree of Animal Development ;
Shewing fundamental Unity in Structure, and the causes of variety; the latter consisting in *Direction* and *Degree* of development.

Fig. 12.

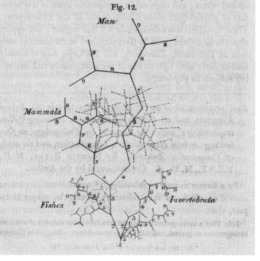

Figure 22. Barry's "Tree of Animal Development" from 1837

specific occurs over nine steps marked with numbers indicating development from order to family, genus, and species and finally to the level of the individual organism. The most general stage, the order, stands at the beginning, while the end of the most delicate branchings represents the individual. As both diagrams show, fish, birds, and mammals—including humans— had stages of development in common. While the majority of English naturalists were still drawing circles, lines, and parallels, hunting for groups of five, or reducing the animal kingdom to dialectical equations, Barry presented the world of animals as a collection of large and small jagged lines that could concentrate into a thicket of marks or spread out in various directions.

In England, the diagrams sent shock waves through the visual world of natural history. Four years later, the research that originated with von Baer and was popularized by Barry had come to be essential knowledge—in 1841 the branching diagram found a place in the second edition of the handbook *Principles of General and Comparative Physiology*, by the physiologist Benjamin William Carpenter (Figure 23). In adapting the diagram Carpenter tidied Barry's tangle of jagged lines into a simple design consisting of one vertical line and four horizontal ones that intersect at right angles only. From the single vertical line, which shows the developmental path shared by all classes of animals, three horizontal lines branch off. The vertical line ends with M (for mammal), passing through the embryonic stages F, R, and B—which stand for fish, reptile, and bird, respectively—on the way. The letters A, B, C, and D to the right of the horizontal lines stand for the final stage of development; D, for example, represents the adult fish. Carpenter's simplification of Barry's diagram had far-reaching consequences: instead of a confusion of lines sprawling out in several directions, the image now showed a step-by-step process of progressive development with a clear orientation from highest to lowest.

Three years later, in 1844, Robert Chambers issued the next version of this image in *Vestiges of Natural History* (Figure 24) to most strongly marked. The view here stated may perhaps receive further elucidation from a simple diagram. Let the vertical line represent the progressive change of type observed in the development of the fœtus, commencing from below. The fœtus of the Fish only advances to the stage F; but it then undergoes a certain change in its progress towards maturity, which is represented by the horizontal line FD. The fœtus of the Reptile passes through the condition which is characteristic of the *fœtal* Fish; and then, stopping short at the grade R, it changes to the perfect Reptile. The same principle applies to Birds and Mammalia; so that A, B, and C,—the *adult* conditions of the higher groups,—are seen to be very different from the *fœtal*, and still more from the *adult*, forms of the lower; whilst between the embryonic forms of all the classes, there is, at certain periods, a very close correspondence, arising from the law of gradual progress from a general to a special condition, already so much dwelt upon. The only exceptions which occur to this statement will be found, from the explanation already given, to be apparent only, and really to prove its truth. (See § 113).

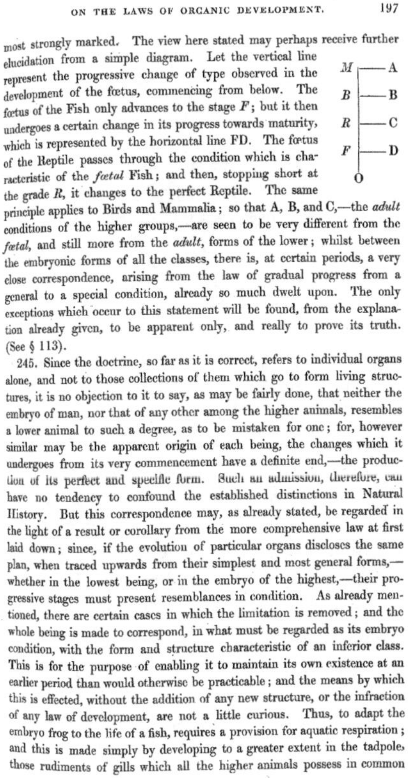

245. Since the doctrine, so far as it is correct, refers to individual organs alone, and not to those collections of them which go to form living structures, it is no objection to it to say, as may be fairly done, that neither the embryo of man, nor that of any other among the higher animals, resembles a lower animal to such a degree, as to be mistaken for one; for, however similar may be the apparent origin of each being, the changes which it undergoes from its very commencement have a definite end,—the production of its perfect and specific form. Such an admission, therefore, can have no tendency to confound the established distinctions in Natural History. But this correspondence may, as already stated, be regarded in the light of a result or corollary from the more comprehensive law at first laid down; since, if the evolution of particular organs discloses the same plan, when traced upwards from their simplest and most general forms,— whether in the lowest being, or in the embryo of the highest,—their progressive stages must present resemblances in condition. As already mentioned, there are certain cases in which the limitation is removed; and the whole being is made to correspond, in what must be regarded as its embryo condition, with the form and structure characteristic of an inferior class. This is for the purpose of enabling it to maintain its own existence at an earlier period than would otherwise be practicable; and the means by which this is effected, without the addition of any new structure, or the infraction of any law of development, are not a little curious. Thus, to adapt the embryo frog to the life of a fish, requires a provision for aquatic respiration; and this is made simply by developing to a greater extent in the tadpole, those rudiments of gills which all the higher animals possess in common

Figure 23. Carpenter's diagram from *Principles of General and Comparative Physiology*, 1841 (Darwin Archive with the permission of the Cambridge University Library Syndicate)

help illustrate the saltation theory of evolution—the theory that presumably prompted Darwin to publish his finch picture in the second edition of *Journal of Researches* a year later. Starting with Carpenter's already modified version of Barry's original, Chambers made two further changes: he tilted the first three horizontal lines upward and completely omitted the fourth. In this way, he gave the ladder of progressive development a dynamic twist. Organisms were striving toward a goal—it seemed

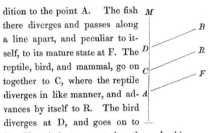

THE VEGETABLE AND ANIMAL KINGDOMS.   219

dition to the point A.   The fish
there diverges and passes along
a line apart, and peculiar to itself, to its mature state at F.   The
reptile, bird, and mammal, go on
together to C, where the reptile
diverges in like manner, and advances by itself to R.   The bird
diverges at D, and goes on to
B.   Here it is apparent that the only thing required for an advance from one grade to another
in the generative process is that, for example, the
fish embryo should not diverge at A, but go on to
C before it diverges, in which case the progeny
will be, not a fish, but a reptile.   To protract the
*straightforward part of the gestation over a small
space* is all that is necessary.

Now we may never see an example of the
working of the actual law which is supposed to
be capable of producing such an advance of
grade ; but something approaching to it in effect
has been observed.   Sex is fully ascertained to
be a matter of development.   All beings are, at
one stage of the embryotic progress, female ; a
certain number of them are afterwards *advanced*
to be of the more powerful sex.   From this it will

L 2

Figure 24. Chambers's diagram from *Natural History of Creation*, 1844 (Darwin Archive with the permission of the Cambridge University Library Syndicate)

as though, deep down, every fish, every reptile, and every bird wanted to be a mammal. In this scheme, fish, reptile, and bird represented developmental byways. The upward flights by which they reached the peak of their development always fell short. Only the mammals stayed on the straight and narrow developmental path to reach the place of honor at the apex of creation. This view of the animal kingdom explains why Chambers believed that the Galápagos islands, which could boast no mammals, were at a low stage of development. Furthermore, Chambers assigned a double function to the diagram: on the one hand, it represented embryonal development, while on the other it showed the historical origins of the orders of animals. The fossil record showed the same steps, from simple, elementary forms to more complex ones that could be observed in embryos as they grew. The deepest levels of the earth's crust contained simple organisms, while fossils became more complex—from fish to reptile to bird to mammal—on the way to the surface. Chambers therefore considered embryonic development to be a model of evolutionary development. He blurred the differences between categories of animals, which most naturalists at the time considered insurmountable, writing, "It is apparent that the only thing required for an advance from one type to another in the generative process is that, for example, the fish embryo should not diverge at A, but go on to C before it diverges, in which case the progeny will be, not a fish, but a reptile." This "law of organic development," as Chambers called it, also applied to the history of the earth: "The idea, then, which I form of the progress of organic life upon the globe is . . . that the simplest and most primitive type, under a law to which that of like-production is subordinate, gave birth to the type next above it, that this again produced the next higher, and so on to the very highest."[57] Chambers thus imagined the process that produced "the type next above" to be a kind of extended pregnancy in which an organism matured to the next stage—the fish, for example, becoming a reptile. "The idea of a Fish passing into a Reptile (his idea) monstrous," Darwin noted in his copy of *Vestiges*.[58]

Even more than Carpenter, Chambers considered the history of living things to be a process of continuous progressive development culminating in the order of mammals. In this regard, both Carpenter and Chambers differed from Barry, who had carefully avoided invoking the concepts of "higher" and "lower," or at least qualified them. Instead of placing the animals in a hierarchy, Barry recommended distinguishing among them in terms of their "degree of elaboration" and "type of structure."

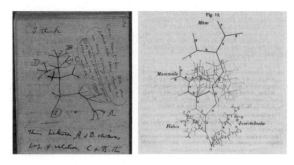

Figure 25. Darwin's and Barry's diagrams from 1837

However, it did not occur to him to derive a theory of evolution based on embryonal development. Instead, he believed that "all finite existences presuppose design."[59] In terms of both individual development and the history of the earth, he viewed the emergence of different types as a product not of evolution but rather of the Creator's defined plan.

Darwin's secret sketch from Notebook B predates the drawings by Carpenter and Chambers and resembles Barry's much more closely, at least at first glance (Figure 25). The same number 1 that Barry had used to mark the starting point in his diagram appeared in Darwin's as well. He likewise divided the image into four sections, which he labeled A, B, C, and D. His sketch also shows a figure with four arms and numerous forks. And both images display the same rampant energy—constant branchings, gaps, and dense thickets of concentrated growth. But these signs took on a different level of meaning in Darwin's version. For Barry, the number 1 referred to Baer's archetype—an imagined unity of organisms or an ideal, divine plan for the animal kingdom. Darwin, in turn, used the number 1 to represent an actual historical origin, the beginning of evolutionary history. The archetype became an ancestor and the principle a starting point.[60]

Like Chambers, Darwin adapted Barry's ontogenetic model into a phylogenetic one, but otherwise the two theories of evolution are nearly diametrically opposed. The visual language employed in each case makes this conceptual distance clear.[61] While Chambers's image of evolution thinned out Barry's drawing to four skinny, orderly lines and arranged the dynamic rays of development into regular angles and steep diagonals, Darwin took the original's profusion to a new level of disorder. In his diagram, lines of development shot out in all directions—sometimes shorter, sometimes longer, without a clear end point or any kind of goal. Furthermore, his drawing incorporated a destructive principle in that it showed not just the development of species, but also their disappearance. Teleology, regularity, and orderliness gave way to accident, variation, and extinction.

Barry's article in the *Edinburgh New Philosophical Journal* thus served as the eye of a needle through which German discoveries on embryology were threaded into the research canon in England. As a result, the question of what kind of natural system underlay the animal kingdom took new form. From now on scientists hoping to discover the ordering principle in nature would have to examine more than the present; determining whether nature was based on a cyclical, binary, linear, or branching structure would require studying its history. By the end of the 1830s anyone who wanted to unearth the system underlying nature had to master the disciplines that examined the development and history of organisms: embryology and paleontology. Ideally such a scientist would be familiar with both the embryonal stages that members of a taxonomic family passed through and the fossil record they had left behind. Ernst Haeckel, a German proponent of evolution, would claim that those who did not possess this level of expertise could bring no more scientific basis to their work than "lovers of any type of artifact" did. With ironic scorn he noted that "systematically arranged collections of coats of arms, . . . of antique furniture, weapons, or costumes, of the so recently popular stamps and other such artifacts can be maintained with just as much sense for specification—with as much joy and interest in the various forms and their systematic grouping—and are often organized and classified with greater logic than collections of snail shells, sea shells, preserved birds, and so on whose owners claim to be 'zoologists.'" The separation between nature and art was valid for both the objects themselves and their admirers: artificial organizing systems existed for hobbyists while zoology was based on genuine genealogical principles. Darwin likewise believed that genealogy was the basis of the only true system of classification. Later, in *Origin of Species,* he would write, "I believe this element of descent is the hidden bond of connexion which naturalists have sought under the term of the Natural System."[62] During the 1840s the belief that the system underlying the animal kingdom could be represented as a geometrical figure became increasingly rare in the taxonomic disciplines. In its place arose the conviction that this system resembled that of a map—just as Strickland had claimed.

PALEONTOLOGICAL DIAGRAMS

The maps first used to chart the history of organisms came from the field of geology. As the seventeenth century drew to a close and explorers and geographers mapped the earth's surface—eventually filling in the outlines of England, Europe, the colonies, and the rest of the world—the inquisitive eye of science bored deeper and deeper into its interior. Following the mines and shafts that led down to resources buried in the bowels of the earth, geologists captured the compact layers below the surface in pictures and opened up previously unimagined realms. They reconstructed extinct plants and animals from fossils and deduced the existence of past epochs and events—volcanic explosions, earthquakes, and floods—from layers of stone. From the fragments they found, they conjured pictures of worlds compressed in stone and compiled overviews in the forms of drawing, diagrams, and maps.[63] Remains of the most amazing organisms were brought up from the earth's depths into the light of day: lizards as big as a house or sea creatures the size of a railway car. While crocodiles or other living exotic animals could be found thousands of miles away, the distance to these

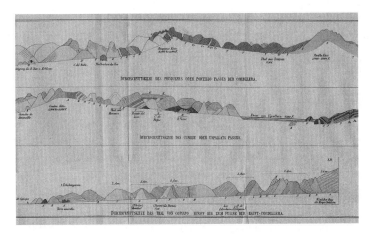

Figure 26. Darwin's cross-section diagram of the Andes after his 1834 expedition (Color Plate 5)

giants was measured in thousands of years. And they shared the earth's interior with yet more animals from even more temporally distant worlds that had to disappear themselves before the later ones could arise. Since time and space correspond in geological maps, each layer down represents an older period. In the face of this vastness, human history shrinks to an almost laughably short span, while the time and space under our feet expand further and further. Darwin never lost his sense of wonder at the scope of the earth's past. "It strikes me," he once commented to a fellow university student, "that all our knowledge about the structure of our Earth is very much what an old hen would know of the hundred-acre field in a corner of which she is scratching."[64]

These dimensions, which Darwin's own cross-section drawings of the interior of the Andes or the Falkland Islands helped to document (Figure 26), were captured by geologists using a symbolic system of lines, points, and dashes. Researchers took what they could learn from a cliff, a mine, or a quarry and combined their findings into a larger view, but these reconstructions required making extensive use of hypotheses. Only by presenting the earth's history visually—compressing time and space into signs—did its duration and physical extent become comprehensible. Geological ages passed between lines just an inch apart on the paper; a horizontal line in a cross-section diagram could indicate a thousand centuries or more, and an entire diagram might symbolize three hundred million years.[65] Like travel guides to the past, these maps illustrated vanished seas, mountains, and forests, along with their inhabitants. In the course of this charting of the earth's interior, Louis Agassiz, who was a student of Cuvier's and later founded the Museum of Comparative Zoology at Harvard, published a series of greatly admired maps that proved to have significant consequences. Like Barry's diagram, they came to serve as models for many images of evolution. His maps' life of their own must have surprised Cuvier, since he later was one of Darwin's staunchest opponents. Unintentionally they encouraged a way of reading that contradicted his own beliefs.[66]

Agassiz published the first fossil map in 1833 while he was still a professor of natural history in Neuchâtel specializing in ichthyology. In the last volume of his work on fossil fish, *Recherches sur les poissons fossiles,* he showed the history of the fish in a map with the title "Généalogie de la classe des poissons" (Figure 27). Following the usual practice with such maps, he placed the earth's layers and ages on the vertical axes and the fish themselves, divided into four classes, on the horizontal. He broke with tradition, however, in the way he showed statistical information on the fossil finds. In the past, researchers had simply provided figures on how often each animal had been found in each geological layer. Agassiz instead chose to use leaflike shapes to show this frequency. Most of the fossils had been discovered where the leaf was widest, and the point where it tapered off marked the point where fossil evidence for the class of fish in question disappeared.[67] This approach made it possible to take in the history of the various classes—their increase, high point, and decline—at a single glance. But the botanical language the

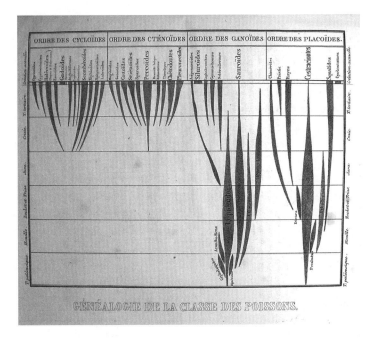

Figure 27. Agassiz's 1833 fossil map from *Recherches sur les poissons fossiles*

illustration cites is contradictory. The long leaves hang disconnected within the earth's history, springing from and fading back into nothing. At certain points it initially seems that one leaf develops from another, but a closer look always reveals a the presence of a narrow channel separating the two. The organic form is missing its most essential characteristics: roots and growth. Agassiz's image consists of frozen foliage that resembles a cut flower more than it does a living plant.

Indeed, this very absence of connections reflects Agassiz's theory of natural history: the lack of a starting point or an evolutionary process meant there was no "trunk" or "roots." Like von Baer and Barry before him, he saw structural similarities between ontogeny and phylogeny. But if embryonal development seemed to echo the earth's larger history, this repetition simply indicated that a common principle ruled both. The steps evident in the fossil record were not the result of a line of descent, but rather evidence of the unfolding of God's millennial plan. Change that occurred during the history of generations was due to the Creator's hand as it modified organisms, introduced new forms, and eliminated others. Following his teacher Cuvier, Agassiz divided the animal kingdom into four *embranchements*—the four fundamentally separate organizational systems known as the *vertebrata, mollusca, articulata,* and *radiata.*[68]

The second chart, issued later, clearly shows this isolation (Figure 28). It indicates the positions of the fossils within a globe. Each class of animals has burrowed its own tunnel into the earth's past without touching any of the others. The narrow separations between the species on the previous map— separations which could so easily be overlooked—have widened here into massive columns. This illustration served as the frontispiece to Agassiz's *Principles of Zoology* (he was now a professor at Harvard), a standard work that had a place in many libraries, including Charles Darwin's. At the center of the image—the middle of the earth—is an isolated black point surrounded by a white circle. The next ring, which fits around the central circle like a gasket, is divided into four segments containing the symbols of the four *embranchements*. No connection exists between the point at the center, insulated in its double envelope, and the fossil layers. As a result, the central point at which the species originate is presented as ideal and immaterial. The fossil record, the earth's layers, and scientific excavations represented a single, contingent side of reality; behind it was a divine conceptual heaven consisting of four domains.

Among those who overlooked the separating lines between the classes of fish in the initial 1833 map—and thus read Agassiz's organic visual language in an evolutionary light—was Charles Darwin. When he wrote in Notebook B that "fish can be traced right down to simple organization," he may already have been referring to the fossil map from *Recherches sur les poissons fossiles;* in *Origin of Species* he referred to it as proof of his

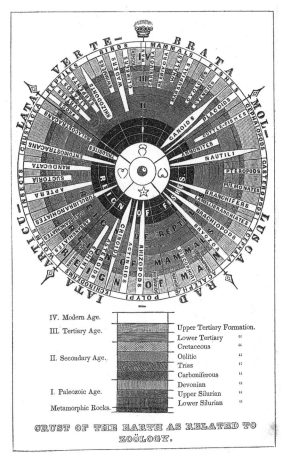

Figure 28. Title page of Agassiz's 1848 *Principles of Zoology* from Darwin's library (Darwin Archive with the permission of the Cambridge University Library Syndicate)

theory. According to Darwin, the map's parallel consideration of the earth's history and embryology in relation to the fishes meant that Agassiz was a witness to Darwin's theory of evolution. He wrote, "Agassiz and several other highly competent judges insist that ancient animals resemble to a certain extent the embryos of recent animals belonging to the same classes; and that the geological succession of extinct forms is nearly parallel with the embryological development of existing forms. This view accords admirably well with our theory."[69] Darwin had already taken the mental step required to turn Agassiz's fish classes into a theory of evolution in the form of a drawing. The archive in Cambridge contains a piece of his blue stationery with a sketch from the early 1850s showing an altered copy of the frontispiece from Agassiz's *Principles of Zoology* (Figure 29). What Agassiz had separated, Darwin's reworked drawing merged: joined by lines of descent, the organisms make their way down through the circle segments symbolizing the earth's history, shown in terms of the "Paleozoic," "Secondary," and "Tertiary Age." But instead of presenting rigid channels, Darwin's drawing shows points spreading out in all directions. "Dot means new form," he noted above the drawing, in which the branching diagram from Notebook B with the caption "I think" reappears before the backdrop of a geological map. The central point has become a point of origin, and the models provided by Barry and Agassiz were fused as in a collage to produce an image that contradicted both of them.

In the years that followed, Darwin completed his picture of evolution, and the sketches became a book illustration. Since 1837 he had tirelessly drawn new diagrams, filling page after page with a luxuriant growth of genealogies, branches, and origins. The sketches, showing the history of rodents or marsupials, became increasingly complex (Figure 30 [Color Plate 6]). Each hypothetical line was tested over the years with countless questions to his correspondents, the study of equally countless journals and books, experiments in his front yard, and statistical analyses. In 1842 he moved from London to a country home in Downe sixteen miles away, and in the same year he produced an initial essay on evolutionary theory consisting of thirty-five densely penciled pages. Since 1844, a further version of his theory of evolution—this one in ink and stretching over two hundred and thirty-one pages—had lain in his closet at home, tied

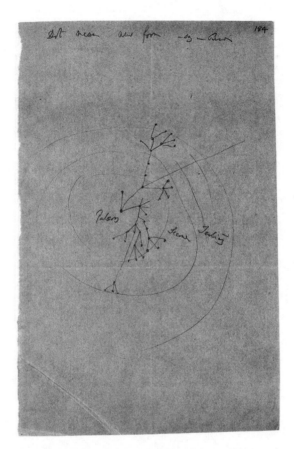

Figure 29. Darwin's diagram from the 1850s based the earlier image by Agassiz (Darwin Archive [DAR 205.5.185], with the permission of the Cambridge University Library Syndicate)

into a packet with ribbon and stored with the tennis rackets. Almost two decades had passed since Darwin made the first sketches in Notebook B. He had married Emma Wedgwood, the daughter of the famous porcelain manufacturer, and had ten children, three of whom died. In 1850s the tenth edition of Robert Chambers's *Vestiges of Natural History* was being published.

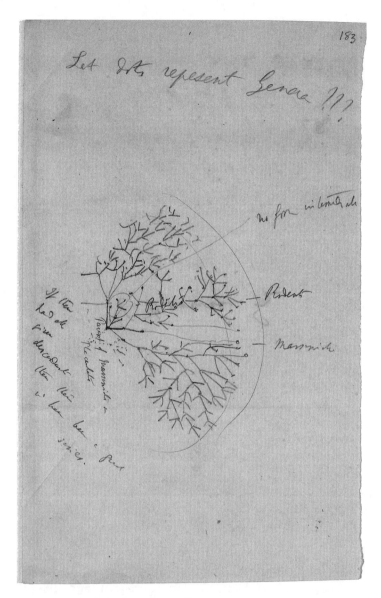

Figure 30. Diagram sketch by Darwin on the rodents and marsupials (Darwin Archive [DAR 205.5.183], with the permission of the Cambridge University Library Syndicate) (Color Plate 6)

During this time, Darwin worked to classify the barnacles—both fossil and living—and published his findings in four volumes. This effort finally proved to him what the Galápagos finches had already shown: that species could split into a countless number of new species and varieties. The Chambers diagram of a step-by-step process of progressive development was wrong—the shape of evolution was that of a tirelessly branching tangle.[70]

DARWIN'S 1859 DIAGRAM IN *THE ORIGIN OF SPECIES*

The event that finally prompted Darwin to abandon his hesitation and publish as quickly as possible is well known: a letter from Ternate, a Malaysian island between Celebes and New Guinea, reached him on a June morning in 1858. It bore the handwriting of Alfred Russel Wallace, an explorer who earned his living by collecting and selling preserved specimens. A year earlier Darwin had asked Wallace to keep an eye out for a rare species of Malaysian bird during his travels. But instead of feathers, the envelope from Wallace contained a summary of a theory of evolution that, to Darwin's horror, resembled his own down to the smallest detail. "I never saw a more striking coincidence," he hastily wrote to Charles Lyell. "If Wallace had my M. S. sketch written out in 1842 he could not have made a better short abstract!" In the accompanying letter Wallace asked Darwin to pass on the text to Charles Lyell if it seemed interesting enough.[71]

In just three years Wallace, who was Darwin's junior by fourteen years, fast-forwarded through the same stages that had led Darwin to the theory of evolution. In 1855 Wallace published *On the Law Which Has Regulated the Introduction of New Species* while in Borneo. This essay drew on Strickland's work and compared the irregular design of the natural system with a branching, gnarled oak or the human circulatory system.[72] The reason for this irregularity, he claimed, was the variation among species, which were forever splitting off into new forms. In 1856

Wallace followed this effort with a comprehensive work on bird classification, *Attempts at a Natural Arrangement of Birds*. Like Strickland, he insisted that "in every systematic work each tribe and family should be illustrated by some such diagram."[73] He also shared the practical steps that needed to be followed to produce such an illustration. In the extensive introduction he recommended using a rifle cleaner to poke small, round markers out of a page, noting the names of the genera on them, sorting the markers along a central axis, transferring the resulting puzzle to a sheet of paper, and connecting the resulting dots with lines (Figure 31). His own diagram created with this method consisted of a long vertical line, along which the names of the families were strung, and horizontal lines branching off at right angles that terminated in the genera. Like Darwin, Wallace claimed that the gaps in the drawing and the distances between some genera within the same family were the result of some species dying out: "All gaps between species, genera, or larger groups are the result of the extinction of species during former epochs of the world's history."[74]

Finally, in August 1858—shortly after Darwin received the letter—Wallace's essay "On the Tendency of Varieties to Depart Indefinitely from the Original Type" appeared, summarizing his previously recorded thoughts on species variation. Thanks to his extensive experience as a collector of and trader in specimens, he could report that varieties and species were impossible to distinguish from one another: "Which is the *variety* and which the original *species*, there is generally no means of determining."[75] One of the sections within the essay states bears the title "Superior Varieties Will Ultimately Extirpate the Original Species."

Darwin had read Wallace's essays, and his fellow researchers, including his now-trusted associate Charles Lyell, had expressed their interest in Wallace's work. But before 1858 Darwin, who was part of England's most influential scientific circles and enjoyed the respect of the age's most renowned scholars, simply could not conceive of an outsider—and one he considered a

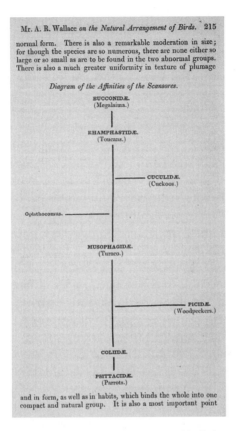

Figure 31. Wallace's diagram of the scansores family of birds from 1858

trader in specimens, at that—becoming a rival. "Nothing very new," he scribbled in the margin of one of Wallace's essays, and "uses my simile of the tree."[76] The 1855 essay Darwin so describes, however, contains the statement "that we have only fragments of this vast system, the stem and main branches being represented by extinct species of which we have no knowledge, while a vast mass of limbs and boughs and minute twigs and scattered leaves is what we have to place in order." Wallace's language here sounds remarkably like Darwin's comment in

Notebook B from 1837: "base of branches dead; so that passages cannot be seen." But Darwin still prematurely dismissed Wallace in another marginal note, writing "it seems all creation with him." Despite this commentary, Wallace followed the same path to reach the theory of evolution that Darwin had: species variation, extinction, the ordering of species within the natural system, the impossibility of distinguishing species from variations, and works of Lyell and Malthus.[77] After consultation with Lyell and Hooker, Darwin's and Wallace's theory of evolution was presented jointly at a meeting of the Linnean Society on July 1, 1858. Twenty-five members were present. Darwin did not attend, and Wallace was still in Asia; in their absence the texts were read out loud. Their presentation was the first of six talks on the meeting's agenda. The absence of the two authors and the fullness of the program both likely played a role in the fact that the theory received scant notice. In the Linnean Society's annual report for 1858, the president remarked that little of note had occurred: "The year which has passed has not, indeed, been marked by any of those striking discoveries which at once revolutionize, so to speak, the department of science on which they bear."[78] Just as Darwin had underestimated Wallace, the society's president had underestimated both Wallace and Darwin.

In November 1859 *Origin of Species* appeared. Darwin worked under tremendous pressure between the summer of 1858 and the fall of 1859 to finish the work, driven by the fear that Wallace could still beat him to the punch. The *Big Species Book,* as he called his original manuscript, was already several hundred pages long but only two-thirds finished. He abandoned this version in favor of a shorter one that ultimately appeared as *Origin of Species.* Of the four diagrams that he had prepared with ink and pasted-on letters for the original book (Figure 32 [Color Plate 7]), only the first appears, with some changes, in the final version.[79] The radial time axis apparent in many of the previous sketches gave way to a horizontal structure; even at this early stage Darwin planned the table as a fold-out insert bound into

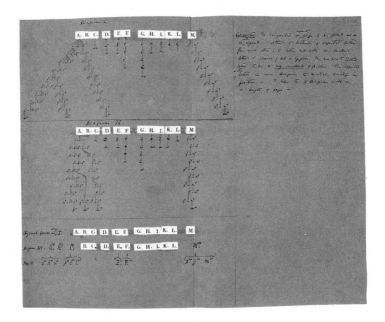

Figure 32. Draft of Darwin's diagram for the *Big Species Book* from 1856–1858 (Darwin Archive [DAR 26rs.opening], with the permission of the Cambridge University Library Syndicate) (Color Plate 7)

the middle of the book for the "convenience of reference" of his still imaginary readers, admonishing them to keep the table at the ready whenever reading the book. Darwin also provided instructions to the engraver, encouraging him to "attend to distance from capital letters to each other."[80] He had placed the letters at irregular intervals and feared that the engraver might "correct" this aspect of his drawings. Unlike the quinarians, whose postscripts record their complaints that the crystalline order and symmetry of their diagrams had not been correctly represented, Darwin was concerned that the irregularities in his table would be lost.

In the final version of the book, the evolutionary diagram appears as a fold-out insert between pages 116 and 117 (Figure 33).

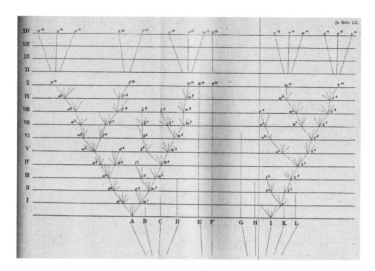

Figure 33. The evolutionary diagram from the 1859 *Origin of Species*

It was both the only picture included in the book and the central image of evolutionary theory. For decades Darwin had worked to capture the panorama of evolution visually, testing and retesting the results so that he could demonstrate the passage of evolution to his readers in a picture. As if the observer could actually see the process of evolution occurring in the diagram, Darwin declared that it illustrated "the probable effects of the action of natural selection through divergence of character and extinction, on the descendants of a common ancestor." In the next paragraph he invited the reader to set this process in motion: "Now let us see how this principle of benefit being derived from divergence of character, combined with the principles of natural selection and of extinction, tends to act."[81] The image begins at the lower margin with a series of points marked A through L placed at irregular distances from one another. This proximity or distance shows the extent to which the species A to L resemble one another. As Strickland had demanded, Darwin reproduces the evolutionary process bit by bit. This panoramic view actually shows the smallest unit of evolutionary history collected in his notebooks so far—one that involved just a few changes. Lines fan out from each point to show the variation of species. Those varieties with characteristics that have proven to be advantageous repeat the process, while the trails of the others break off.

In some cases the history of a species follows a straight line, but most branch off in other directions. Within the very first leg of the evolutionary journey, between the first two horizontal lines, the interplay of selection and variation is already evident, and as this interaction travels along the y-axis of time it traces evolution's zigzag path. Each branch represents an attack on the balance of the natural economy. Stage for stage, variation causes the species not just to grow different from the previous generation, but to encroach into the niches of neighboring species. Variation flourishes within a species at the expense of those surrounding it. As Darwin suggests, competition is "most severe between allied forms, which fill nearly the same place in the economy of nature."[82] The line indicating varieties of B ends at the point where line A widens out. At all the stages that follow, this process of variation, selection, and the extinction of neighboring species repeats, building the process of evolution. What Darwin describes in linear form and in great detail over the course of the book's five hundred pages can be seen at once in all its multilayered simultaneity in the diagram: the variation, branching out, and extinction of species.

Unlike the previous diagrams—the drawings from Notebook B or the radial sketch from the 1850s—the published image omits the point of origin. Darwin shows an evolutionary excerpt, a close-up of a process framed above and below by far greater periods of time that he indicated with rough dotted lines. The uppermost section, which shows the process continuing infinitely, reminds readers that evolution has not come to a stop upon reaching the top of the page. The present is just a temporary state as well. In turn, the lines that trail away at angles from the row of points to the bottom of the page show the events that preceded this excerpt—evolution's endless history. If the lines continued, those that incline slightly toward one another would eventually meet. This hint is the closest Darwin comes to providing a starting point, leaving his reader to follow it and resolve the paradox of the "origin" invoked in the title but missing from the book.[83]

Darwin not only hesitated to indicate a point of origin in his diagram, he also avoided giving his picture of evolution a clear form. To his frustration, he discovered that symmetry, order, and regularity took form on paper much more quickly than they did in nature. He therefore apologized for the impression of regularity that standardized printing processes forced upon the printed versions of his diagrams, in contrast to the drawings he had done by hand. "But I must here remark," he interjected after the first explanatory paragraphs, "that I do not suppose that the process ever goes on so regularly as is represented in the diagram, though in itself made somewhat irregular."[84] The irregularity he had achieved in his drawings appeared only to a limited extent in the printed diagrams. It was apparently even more difficult for Darwin to capture in words what the picture shows; a total of eleven pages of text are needed to elucidate this single-page illustration. A type of viewing guide scattered over three different passages explains the meaning of the horizontal lines, diagonals, marks, Roman numerals, and letters. The first and longest appears in the fourth chapter, "Natural Selection," while the other two, far shorter, form part of Chapter 14, "Classification." The first reference to the chart occurs as his argumentation reaches a turning point. Darwin has explained the two cornerstones of his theory of evolution in the preceding chapters: first, that species vary from generation to generation and that these variations can be advantageous or disadvantageous to the offspring; and second, that nature's resources are limited, meaning that the better-equipped variety will eliminate the less advantaged in the "struggle for existence." Darwin illustrates the principles of variation, selection, and struggle using examples of living species. He then introduces the diagram at

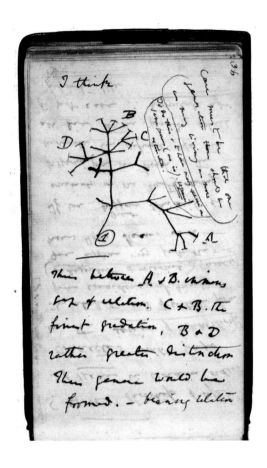

Plate 1. Darwin's 1837 diagram from Notebook B (Darwin Archive [Notebook B], with the permission of the Cambridge University Library Syndicate) (Also Figure 12)

Plate 2. A Galápagos finch with a museum label filled out by
Darwin (With the permission of the Natural History Museum,
London) (Also Figure 13)

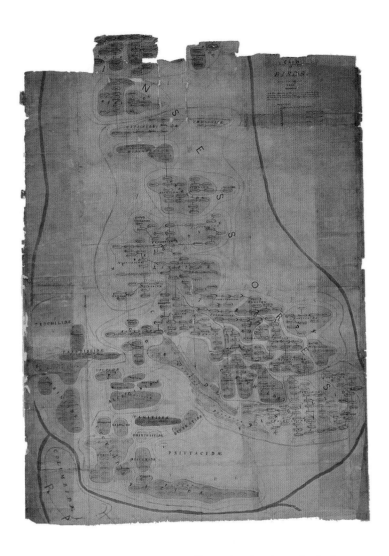

Plate 3. Strickland's 1840 wall map for classifying the birds
(With the permission of the Zoological Museum, Cambridge)
(Also Figure 15)

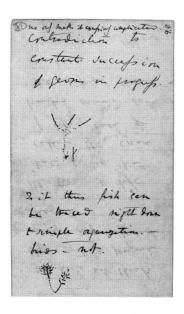
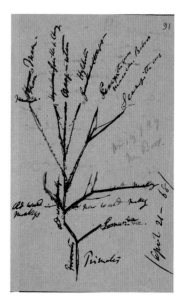

Plate 4a. Two diagrams in Notebook B occurring a few pages before the "I think" sketch (Darwin Archive [Notebook B], with the permission of the Cambridge University Library Syndicate) (Also Figure 19)

Plate 4b. Darwin's 1868 family tree of the primates (Darwin Archive [DAR 84.91], with the permission of the Cambridge University Library Syndicate) (Also Figure 54)

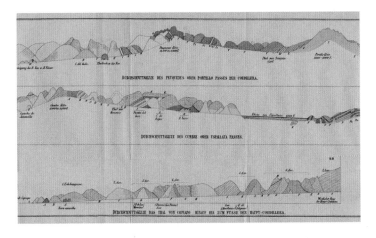

Plate 5. Darwin's cross-section diagram of the Andes after his 1834 expedition (Also Figure 26)

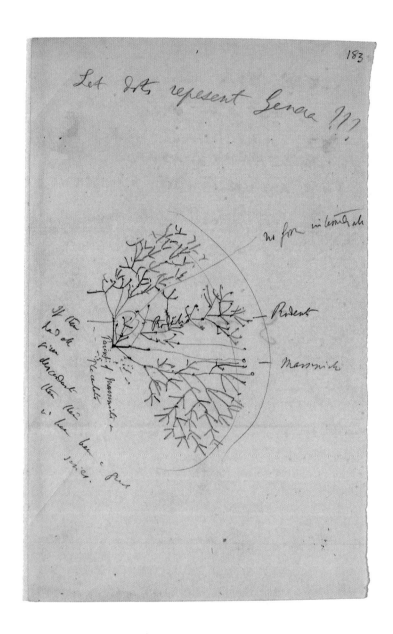

Plate 6. Diagram sketch by Darwin on the rodents and marsupials (Darwin Archive [DAR 205.5.183], with the permission of the Cambridge University Library Syndicate) (Also Figure 30)

on fossil findings. The family trees prepared by Ernst Haeckel, who began corresponding with Darwin in 1863, became particularly famous.

Like Darwin, who was twenty-five years his senior, the German Haeckel initially studied medicine, but after receiving his doctoral degree he decided to pursue a *Habilitation*—an academic degree beyond the doctorate in Germany—in comparative anatomy. Haeckel taught evolutionary theory as a private lecturer at the university in Jena starting in 1862; his 1866 book *Generelle Morphologie der Organismen* (General Morphology of Organisms) offered, as he claimed, the first concrete implementation of Darwin's theories in taxonomy.[86] While he referred to the diagrams in this two-volume work as "family trees," they more closely resemble the "shrubs" which Strickland had invoked to describe the natural system (Figure 34). Contortions were required to make the reconstructions of animal and plant evolution fit within the page's boundaries. On each of the eight illustrations, the tree splits into several branches just above the roots, which then divide in turn again and again. On one plate Haeckel echoes Agassiz's fossil map from the 1833 *Recherches sur les poissons fossiles*, citing the organic visual language used here to join all divisions of the "vertebrate family tree" in connected feeders. Like a vine the branches grow from the trunk at the lower left across the page to the right.[87] The plate shown here, in turn, depicts the "mammal family tree." In this jumble of branches, one upper twig sends out a shoot that produces a second bush emerging from the first at the point where *Zonoplacentalia* and *Discoplacentalia* fork off in different directions. Haeckel tucked away humans in the section devoted to anthropoids in the upper right corner of the page, side by side with the gorillas and at the same height as the order *felina*, the cats.[88] Upon reading Haeckel's *Generelle Morphologie* in 1866, Darwin felt that his image of evolution was in good hands: "My dear Haeckel," Darwin wrote, "Your boldness sometimes makes me tremble but . . . someone must be bold enough to make a beginning in drawing up tables of descent."[89]

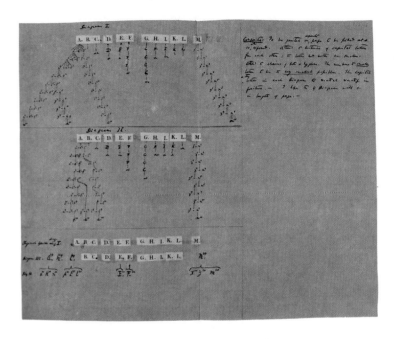

Plate 7. Draft of Darwin's diagram for the *Big Species Book* from 1856–1858 (Darwin Archive [DAR 26rs.opening], with the permission of the Cambridge University Library Syndicate) (Also Figure 32)

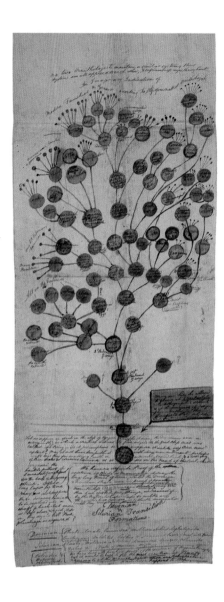

Plate 8. Family tree of the birds in an 1861 letter from one of Darwin's correspondents (Darwin Archive [DAR 171.56], with the permission of the Cambridge University Library Syndicate)
(Also Figure 36)

the point where these principles join to produce tl
action of the historical evolutionary process. The ii
an opportunity for simultaneously showing extinct
plosive growth, variation and competition, a detai
overall view, and history and the present moment ir
the text used to explain it does not. The passages t
the diagram line by line and letter by letter, dissectii
play of the shapes on the page, are laborious rea
language must proceed from one detail to the next,
covers a period of time far too long to be observe(
More than "306,662,400 years; or say three hund
years"[85] fit on this single page, where a thousand ger
no more than a square inch, to demonstrate the cre
tial of chance variation in hosts of tiny marks. Chan
and selection, which the observer can see only in
form in nature, are joined in this systematic view of
action. The cumulative impact of small factors ovei
years is shown in the slant of a line.

Those who might have found the greatest satisfa
erratic zigzag path of the natural system had die(
book's publication. Hugh Strickland, whose essay '
Method of Discovering the Natural System in ^
Botany" opened the hunt for the unknown masterj
the natural system, was run over by a train in 18^
lecting fossils between the tracks. The man Darwir
as one of the executors of his scientific papers th
before Darwin himself, and natural history went i
out him. Darwin went on to win the race with a
combined three disciplines like a collage: Barry
diagram took on overgrown proportions as it met
taxonomic maps and was then planted deep in
substrata. With his 1859 diagram Darwin creat
print for the so-called family trees that evolutiona
still prepare today. He offered a general model
that could be filled in with the concrete developm
cies, genus, family, order, or the entire animal kii

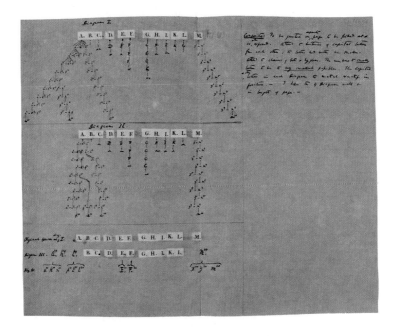

Plate 7. Draft of Darwin's diagram for the *Big Species Book*
from 1856–1858 (Darwin Archive [DAR 26rs.opening], with the
permission of the Cambridge University Library Syndicate)
(Also Figure 32)

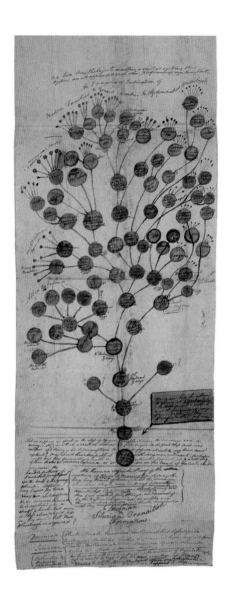

Plate 8. Family tree of the birds in an 1861 letter from one of Darwin's correspondents (Darwin Archive [DAR 171.56], with the permission of the Cambridge University Library Syndicate) (Also Figure 36)

the point where these principles join to produce the dynamic action of the historical evolutionary process. The image offers an opportunity for simultaneously showing extinction and explosive growth, variation and competition, a detailed and an overall view, and history and the present moment in a way that the text used to explain it does not. The passages that explain the diagram line by line and letter by letter, dissecting the interplay of the shapes on the page, are laborious reading. While language must proceed from one detail to the next, the picture covers a period of time far too long to be observed in nature. More than "306,662,400 years; or say three hundred million years"[85] fit on this single page, where a thousand generations fill no more than a square inch, to demonstrate the creative potential of chance variation in hosts of tiny marks. Chance variation and selection, which the observer can see only in fragmented form in nature, are joined in this systematic view of evolution in action. The cumulative impact of small factors over millions of years is shown in the slant of a line.

Those who might have found the greatest satisfaction in this erratic zigzag path of the natural system had died before the book's publication. Hugh Strickland, whose essay "On the True Method of Discovering the Natural System in Zoology and Botany" opened the hunt for the unknown masterpiece behind the natural system, was run over by a train in 1853 while collecting fossils between the tracks. The man Darwin had named as one of the executors of his scientific papers thus died long before Darwin himself, and natural history went its way without him. Darwin went on to win the race with a picture that combined three disciplines like a collage: Barry's branching diagram took on overgrown proportions as it met Strickland's taxonomic maps and was then planted deep in the geologic substrata. With his 1859 diagram Darwin created the blueprint for the so-called family trees that evolutionary biologists still prepare today. He offered a general model of evolution that could be filled in with the concrete development of a species, genus, family, order, or the entire animal kingdom based on fossil findings. The family trees prepared by Ernst Haeckel, who began corresponding with Darwin in 1863, became particularly famous.

Like Darwin, who was twenty-five years his senior, the German Haeckel initially studied medicine, but after receiving his doctoral degree he decided to pursue a *Habilitation*—an academic degree beyond the doctorate in Germany—in comparative anatomy. Haeckel taught evolutionary theory as a private lecturer at the university in Jena starting in 1862; his 1866 book *Generelle Morphologie der Organismen* (General Morphology of Organisms) offered, as he claimed, the first concrete implementation of Darwin's theories in taxonomy.[86] While he referred to the diagrams in this two-volume work as "family trees," they more closely resemble the "shrubs" which Strickland had invoked to describe the natural system (Figure 34). Contortions were required to make the reconstructions of animal and plant evolution fit within the page's boundaries. On each of the eight illustrations, the tree splits into several branches just above the roots, which then divide in turn again and again. On one plate Haeckel echoes Agassiz's fossil map from the 1833 *Recherches sur les poissons fossiles*, citing the organic visual language used there to join all divisions of the "vertebrate family tree" in connected feeders. Like a vine the branches grow from the trunk at the lower left across the page to the right.[87] The plate shown here, in turn, depicts the "mammal family tree." In this jumble of branches, one upper twig sends out a shoot that produces a second bush emerging from the first at the point where *Zonoplacentalia* and *Discoplacentalia* fork off in different directions. Haeckel tucked away humans in the section devoted to anthropoids in the upper right corner of the page, side by side with the gorillas and at the same height as the order *felina*, the cats.[88] Upon reading Haeckel's *Generelle Morphologie* in 1866, Darwin felt that his image of evolution was in good hands: "My dear Haeckel," Darwin wrote, "Your boldness sometimes makes me tremble but . . . someone must be bold enough to make a beginning in drawing up tables of descent."[89]

Figure 34. Haeckel's "family tree of the mammals, including man," from 1866

After 1859 the standard practice was to depict evolutionary time using images that show how constant small changes, over long periods, accumulate to become large ones. But Darwin's visualized disorder was lost. In terms of their reception, the pictures he had created—with paper, ink, and pen and using the elements of point, line, and circle—were apparently so rich in relationships that they practically demanded to be simplified. The images that came after Darwin reversed the paradoxical quality of his illustrations: the seemingly simple visual language

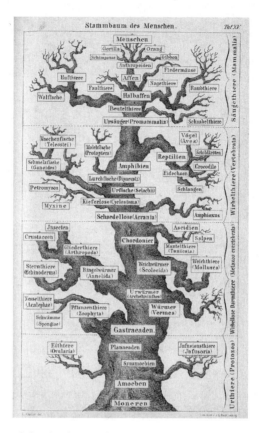

Figure 35. Haeckel's "family tree of man," from 1874

of artless pictorial elements he used to express his multilayered theories gave way to representations that used far more elaborate forms, but simplified how evolution was portrayed in theoretical terms. Haeckel's ornate visual language, which forced the evolutionary undergrowth he initially conceived into the shape of a tree, achieved a prominence that was not without troubling consequences (Figure 35). The luxuriant oak from his 1874 book *Anthropogenie, oder Entwickelungsgeschichte des Menschen* (Anthropogenie, Or the Evolution of Man) became his most famous picture and defined the appearance of the evolutionary family tree for many years. In this image, humans were enthroned at the highest point of the crown. From the right-hand corner of the 1866 version of this evolutionary diagram—where human, gorilla, or cat could find itself at the leaf's upper tip—a center and peak had emerged.[90] Unlike Darwin, whose fold-out table showed evolution to be an ongoing process, Haeckel brought evolution to a halt once it reached mankind. The oak was mature and its crown fully developed. Evolutionary history became a story of progress through upward-leading stages. The principles of branching and continuation disappeared almost completely in this tree, whose powerful trunk concealed a self-enclosed stepladder at its heart. The decoration provided by the branches and leaves recalls Chambers's diagram from *Vestiges of the Natural History of Creation* (see Figure 24), which portrayed the history of living things as a constant development into higher forms back in 1844.[91]

Many of Darwin's contemporaries doubted that the integration of chance and disorder as Darwin envisioned it had the makings of a good scientific theory. For example, John Mivart, whose initial fiery enthusiasm and later bitter opposition made him infamous as the "fallen angel" of the evolutionary movement, compared evolutionary history with a rolling polyhedron: he claimed that the system of nature formed a symmetrical, harmonic whole at all times, and that change was purely due to the polyhedron changing position and coming to rest on a different surface.[92] Even Owen, who had classified the fossils from the *Beagle* voyage, published a visual retort to Darwin's diagram in 1860. His *Paleontology: Or a Systematic Summary of Extinct Animals and Their Geological Relations* includes a horizontally layered line diagram in which he noted the fossils of the mammals. In the place of Darwin's linked trail of living beings, the viewer was confronted with isolated entries. Every triangle and diamond in Owen's picture stood for a class of mammals, which was not connected to any other.[93] The geometrical visual language of Mivart and Owen reveals an anxiety that went public in an infamous joke by John Herschel. Herschel, who in 1830 had declared taxonomy to be the main task of natural history, called Darwin's theory the "law of the higgledy-pigglety." For the astronomer Herschel, schooled in the mathematically comprehensible laws of inorganic nature, evolution that produces disorder was hardly a satisfactory natural principle.[94]

Darwin's suggested solution contradicted paradigms of both natural history and the physical sciences. On the one hand, natural selection was a mechanistic answer to the question of species development that broke with the basic conviction of naturalists: that at the level of the organism or of species, development requires teleology. From Kant to von Baer, from Whewell to Agassiz—they all believed that neither organisms nor species developed randomly, but rather that they followed formative laws that unfolded within the context of a goal-directed existence.[95] Darwin broke with this teleological conception of nature; according to his theory, organisms varied by chance, and the history of living things can take random turns as the countless small variations preserved by selection accumulate. Darwin's evolution did not follow a goal. But neither did the mechanisms of variation and selection he described constitute a law comparable to the classical laws of mechanics, which could be expressed in mathematical terms. As a result, he did not prove to be the "Newton of a blade of grass." Instead, he wrote that he did not believe in any "fixed law of development"; his theory of evolution described an unforeseeable process, a hodgepodge of becoming and passing away whose progress, while indeed determined by mechanisms, was not governed by laws.[96]

Ironically, Herschel's talk of "the law of the higgledy-pigglety" that so offended Darwin seems to capture the special quality of his theory. Comparable labels in art history—from *gothic* to *impressionism*—came to describe their epochs despite their original pejorative intent. The title "law of the higgledy-pigglety" would also have been appropriate for the minor masterpiece that one of Darwin's correspondents presented to him shortly after the publication of *Origin of Species* (Figure 36 [Color Plate 8]).[97]

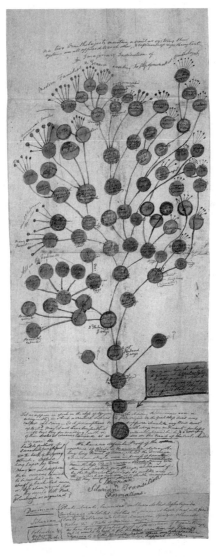

Figure 36. Family tree of the birds in an 1861 letter from one of Darwin's correspondents (Darwin Archive [DAR 171.56], with the permission of the Cambridge University Library Syndicate) (Color Plate 8)

Following Darwin's diagram of evolution, its artist showed—like Swainson, Strickland, and Wallace before him—the order of birds. As if he were following Wallace's arts-and-crafts instructions, the correspondent had used a circular cutout to draw a ring for each species and connected them with lines. From their starting point, the circles multiplied across the page, wound their way into all four corners, and crowded together to the point that they seemed to trip over one another. From each circle emerges another tiny crown from whose points new circles grow, until the edge of the page brings evolutionary history to a temporary halt. The image, the author told Darwin apologetically, was just "a mere rough sketch (on rough paper)."[98] But it was precisely these traces of the handmade that best captured the nature of evolutionary theory. The correspondent's jaunty ink lines and circles on a piece of cardboard recall the discovery of disorder that Darwin ushered into our picture of nature.

3 THE PICTURE SERIES

On the Evolution of Imperfection

IN 1867, eight years after his evolutionary diagram was published in *Origin of Species,* Darwin sat once again at his desk and drew. He developed several pictures at the same time, all displaying his typical awkward artistic style. A few years later, professional artists produced more polished versions of the drawings, which finally appeared as a series of woodcuts in *Descent of Man.* Darwin prepared these drafts on the same paper he had used during previous years. In the period up to 1859, the identical light-blue surface had repeatedly formed the background on which he spread out the panorama of evolution. Now the compositions he sketched in ink and lead pencil seemed more abstract, and he added extensive notes and, in some cases, an index (Figures 37, 38, and 39 [Color Plates 9, 10, 11a]). But in contrast to the diagram from *Origin of Species,* these sketches did not represent the draft of a theory: no nonfigurative symbols representing scientific phenomena, such as bent lines to show variation, were used. Instead, the subject of these three pages, unpublished until now, was a specimen from the British Museum (Figure 40 [Color Plate 11b]). In highly abstracted form, the sketches showed the pattern that the argus pheasant—an Asian member of the pheasant family—sports on the feathers of its wings. Each drawing features a section of a single feather

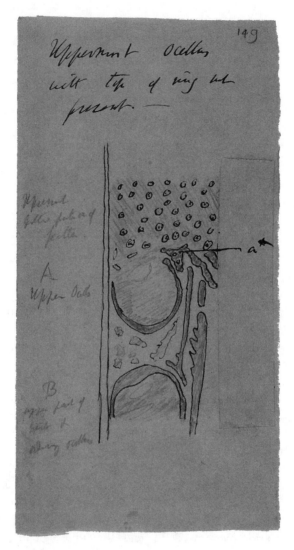

Figure 37. First of Darwin's three sketches of the argus pheasant's feather, circa 1867 (Darwin Archive [DAR 84.149], with the permission of the Cambridge University Library Syndicate) (Color Plate 9)

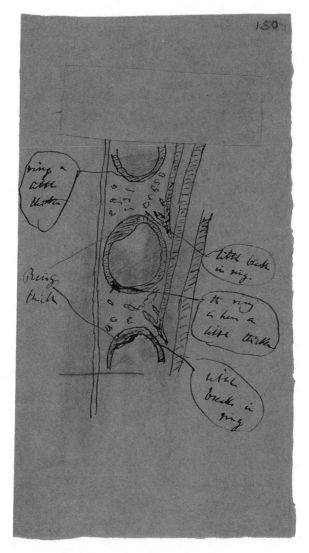

Figure 38. Second sketch of the argus pheasant's feather (Darwin Archive [DAR 84.150], with the permission of the Cambridge University Library Syndicate) (Color Plate 10)

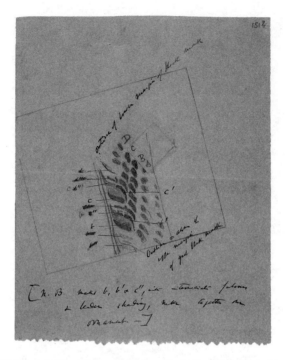

Figure 39. Third sketch of the argus pheasant's feather (Darwin Archive [DAR 84.151b], with the permission of the Cambridge University Library Syndicate) (Color Plate 11a)

with a different variation of the eye-shaped ornament, or "ocellus," that Darwin found on the stuffed bird before him and copied vertically along the quill. Since the birds he examined were fully grown males, the drawings reproduce the ornaments in its real-life proportions, with each eye measuring about an inch across.

Taken as drawings meant to represent a real object, Darwin's sketches seem unsuccessful at first glance. In all three cases, the subtle color transitions and indistinct patterns of the original are rendered in high-contrast black-and-white and the broadest types of geometrical forms, their irregular outlines taking shape as comblike protrusions, dots, whorls, waves, or lines. The pic-

Figure 40. Argus pheasant's wing from the former collection of the British Museum from the 1860s (Color Plate 11b)

ture created by nature and its reproduction are thus at odds with one another. While nature decorated the wings of the argus pheasant in a trompe l'oeil style that, as one of Darwin's contemporaries aptly said, resembled a ball in a socket, the image on the paper is an irregular, abstract collection of forms. Nature seems more artistic here than its artificial copy. But a second look shows that the seemingly failed drawings evidence signs of astounding precision. All four sheets contain areas that have been pasted over with slips of paper and redrawn—apparently to fix mistakes. Furthermore, written comments address those shortcomings in the drawings that could not be pasted over. "Rings a little thicker" appears on the second page, like a thought balloon, at the end of a line leading to the edge of a circular segment (see Figure 38). "With breaks in ring" another comment reads, admonishing the observer to introduce gaps mentally into the continuous line shown in the drawing.[1]

A perusal of *Descent of Man* provides the reason why Darwin spent so much effort clarifying the details of his sketches, even at the points where his drawing talent failed him. In the archive, the sketches form part of a chaos of preparatory notes, drawings, and test prints for the book, all dated between 1867 and 1871.[2] The sketches reappear as woodcuts in the published

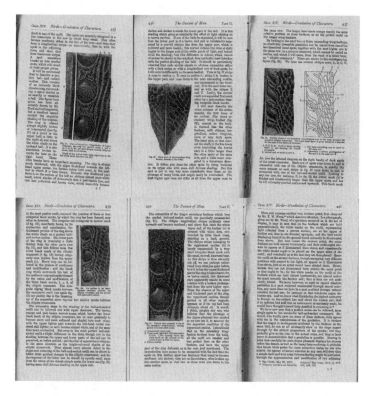

Figure 41. Pages from Darwin's *Descent of Man* containing the series of pictures on the argus pheasant

book in 1871, prepared by George Henry Ford, an artist at the British Museum who otherwise specialized in reptiles and fish (Figure 41).[3] Although Ford's woodcuts surpass their models in terms of craftsmanship—the lines are steady, the shading more nuanced, the details more precise—the most significant characteristics of the sketches remain. Like Darwin, Ford has also reproduced the ornament in black and white rather than in color, and the focus in both cases is on individual, geometric, abstract figures that combine to form the feather pattern as a whole and that are keyed to an index.

But while Darwin's preliminary drawings survive in the archive as a collection of individual pages, the final images based on them are now part of a series of images. Like a flip book scattered over several pages, they show an increasingly complex pattern taking shape on the plumage of the argus pheasant. Page after page, Darwin makes the reader into a witness of an evolutionary process by confronting him or her in a series of pictures with its stages. A sketch showing blotches reappears as Darwin's Figure 58, the second in the series, which depicts the ornament in its simple original form. A sketch with circles, in turn, shows up as Figure 61. This final image in the series presents the pattern in its most developed stage.[4] The five pages thus reconstruct the possible path the ornament followed as it took shape, in which each step—as Darwin emphasizes—is represented with a picture of a specific example he found on the bird's plumage: "In order to discover how the ocelli have been developed, we cannot look to a long line of progenitors. . . . But fortunately the several feathers on the wing suffice to give us a clue to the problem, and they prove to demonstration that a gradation is at least possible from a mere spot to a finished ball-and-socket ocellus."[5] Insofar as the series of pictures drawn by Darwin shows the possible course of evolutionary events, it represents a process that occurs too slowly to be observed in nature. Beyond such a representation, we cannot perceive the history of the pattern: fossilized remains of an ornament on a bird's plumage exist no more than do observational data for a process lasting thousands of generations.

At first glance, Darwin's picture series raises more questions than it answers. The most obvious of these concern the title of the book, as *Descent of Man* includes more illustrations of the argus pheasant than of any other subject. Of the book's seventy-eight illustrations, ten show different ornamented birds, and five of these depict the evolution of the elaborate pattern on the argus pheasant's wings in painstaking detail.[6] Dashing the reader's expectations for an examination of the descent of humankind, however, the book offers not a single picture of a human being, with the exception of a single image of an embryo. Furthermore, although Darwin never tried to reconstruct the evolution of any particular animal or plant in his works—whether *Origin of Species, Variation Under Domestication,* or his monographs on orchids, vines, or earthworms—he lavished attention in this single case on a single object: the argus pheasant's plumage. The feathers thus offer a startling exception within an oeuvre otherwise free of reconstructions. While he provides countless examples of organs or organisms that show traces of evolution, or suggests generally that, for example, humans must have apelike ancestors, he leaves open the precise evolutionary steps involved. This quality caused Sternberger aptly to remark that, in Darwin's works, the "'imperceptible gradations,' by the way, always remain pallid."[7] The feathers of the argus pheasant thus do not just occupy an unexpectedly large place in a book on the history of humanity, but in Darwin's work as a whole.

Darwin's choice of the argus pheasant's feathers as the medium for showing the steps of the evolutionary process in all their hypothetical detail initially seems deeply puzzling. The motif of his pictorial series significantly differs from those found later in schoolbooks about evolutionary theory: the apelike creatures that stand up, step by step, to become *Homo sapiens,* the fish that becomes a lizard, the dinosaur that turns into a bird, or even just the evolution of a three-toed primitive horse into the true ungulates of today.[8] But to explain the principle of step-by-step evolution, Darwin chose the feather of the argus pheasant as an example. Why? The argus pheasant pictures are thus noteworthy in two respects: first, the object they show, and second, how the various pictures line up to form a series that depicts an evolutionary process. Even today, any textbook or work of biology based on evolutionary theory includes such sequences of pictures to show the step-by-step changes in a species, a feature, an organism, or an organ.[9] The medium of the picture series has thus proven to be essential to the history of evolutionary theory, while the argus pheasant as a subject has not. The generations of scientists that followed apparently felt that the bird was less representative than Darwin himself did, and it has reentered the scientific discussion only recently.[10]

In keeping with these two perspectives, this chapter consists of two parts. The first examines the history and function of the picture series, while the second looks at Darwin's choice of subject matter—in other words, the circumstances under which the argus pheasant could become central to the theory of evolution. During years of research, Darwin followed both strands to spin an ever-growing web of notes, observations, drawings, models, pictures, and letters—a web that stretched back historically into the eighteenth century and extended to encompass correspondents, explorers, dealers in animals, and colonial officials from Downe in County Kent to the tropics of Malaysia.[11] As he drew this net together, old connections at a number of levels unraveled and new ones came into being. In the end, the human view of the bird's striking plumage seemed as foreign to him as the perspective that an "inhabitant of another planet"[12] would have on humans themselves. The human eye thus lost the central position it had commanded for centuries as the perspective from which all things were to be observed. It was displaced by that of the argus pheasant hen, for whose admiring gaze the male pheasant fans his wings during the mating dance. This change in perspective forced the human observer to abandon any belief in nature's perfection and status as an artwork. Darwin's pictures revealed an aesthetic based on change and showed the mutability of a world eternally in progress.

THE PREHISTORY OF THE PICTURE SERIES

By including a picture series in *Descent of Man* in 1871, Darwin drew on a type of image that he had seen in many forms in the books in his library, where it appeared as lithographic inserts, fold-out plates, copper etchings on title pages, or woodcuts interspersed within the text. Just as geology ushered in new types of images when it emerged as a science at the end of the eighteenth century, three young disciplines during this time now placed particular focus on the picture series: embryology, physiognomy, and comparative anatomy. However, despite the formal similarity in each case—the arrangement of several illustrations in a sequence—the picture series fulfilled very different purposes in each field. In comparative anatomy, for example, it allowed scientists to study a variety of organisms in terms of their structural similarities and differences. From Pierre Belon's original presentation of a human and a bird skeleton in the 1555 *L'histoire de la nature des oyseaux*, the number of organisms involved in such comparisons grew to three or more. In 1805 Cuvier's last volume of *Leçons d'anatomie comparée* included images of the limbs of a monkey, cat, bear, sea lion, and dolphin next to one another, arranging the plate's ten illustrations in vertical and horizontal rows. All the extremities are presented in the same size, so that the cat's paw and the monkey's hand are just as big as the bear's paw and the dolphin's fin. Unlike actual specimens in an anatomical collection, pictures could make comparison easier by standardizing the size of the objects, the section shown, and the angle of view.[13]

Similarly, Richard Owen—Darwin's former paleontologist for the *Beagle* collection, coiner of the word *dinosaur* (literally "terrible lizard"), and now professor at the Hunterian Museum of the Royal College of Surgeons—presented the skeletons of various vertebrates together on a single page in 1848 (Figure 42).[14] In his *On the Archetype and Homologies of the Vertebrate Skeleton* and *On the Nature of Limbs* he showed side views of human and bird skeletons on the left half of a large fold-out plate and those of a fish, reptile, and mammal on the right half. Read from top to bottom, the series demonstrates the "archetype" or structural plan of the vertebrates that, in Owen's theory, underlies all forms that animals within this zoological category could take. To construct this archetype, he divided the vertebrate skeleton into sections and argued that similar structures recur from animal to animal. The same anatomical elements would appear with slight changes, depending on the environment. These comparative methods allowed him to abstract common structures

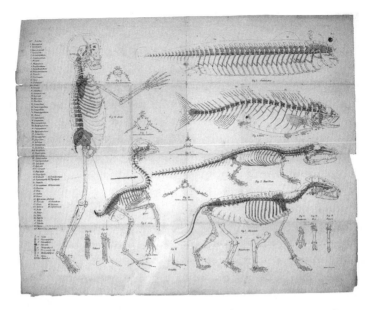

Figure 42. Owen's table on invertebrate anatomy from 1848 (Darwin Archive with the permission of the Cambridge University Library Syndicate)

from the multiplicity of forms actual animals exhibit. His readers could see the reasons why Owen believed all vertebrates shared a common structural plan for themselves by comparing the archetype with the other skeletons on the fold-out.

Although Owen conceived of the archetype as a type of blueprint, he also indicated that this abstract form closely resembled the fossils of certain early types of vertebrates. "The Archetypal idea was manifested in the flesh," he wrote, echoing the biblical account of the word that becomes flesh, "under divers such modifications, upon this planet, long prior to the existence of those animal species that actually exemplify it." Four years after Chambers's *Vestiges* appeared, Owen was fully aware of how easily the sequence in which he arranged the vertebrates for comparison in his table could be seen as representing steps in a process of development. Following the logic of Chambers's

diagram, in which fish, reptile, bird, and mammal were developmental stages, it was also tempting to read the order of Owen's
skeletons as a temporal sequence. He did not encourage this interpretation, however, claiming that "to what natural laws or secondary causes the orderly succession and progression of such
organic phaenomena may have been committed we as yet are
ignorant."[15] He thus left unresolved the question of whether the
table showed a chain of independently created beings or an inter-
connected sequence. It was still possible to imagine that the archetype was like a common floor plan for buildings that resemble
one another without actually producing each other through a
natural process.

The science of physiognomy made just such a use of the picture series. Darwin's library contains a heavily annotated ten-
volume French edition of Johann Caspar Lavater's *Physiognomic
Fragments for the Promotion of the Knowledge and Love of Mankind*. This expanded version of the original German treatise,
which was published between 1775 and 1778, had been autho-
rized by the Swiss pastor, philosopher, and author. The most
striking feature of the French edition was a new illustrated
plate. Based on a sketch of Lavater's, it showed a twenty-four-
step transformation from frog to Greek god (Figure 43).[16] Obviously, the sequence of profiles from frog to Apollo does not represent a temporal process. The hybrid man-frogs, or frogmen,
do not exist now and never did, and the author does not assume
them to be real. Instead, the series of drawings offers Lavater a
way to demonstrate the principles on which human and animal
faces are constructed. The intermediary steps are meant to show
how a number of details combine in our perception of the ugly
and the beautiful.[17]

Unlike the fields of physiognomy and comparative anatomy,
the third discipline in which series of images played a decisive
role—embryology—did use this type of picture to show a process
in time. One of the first scholars to show human embryological
development in its entirety was Samuel Thomas von Soemmering, a naturalist who corresponded with Goethe. His images

Figure 43. Lavater's 1809 physiognomic comparison of profiles from Darwin's library (Darwin Archive with the permission of the Cambridge University Library Syndicate)

appeared in a well-known pictorial atlas in 1799 (Figure 44) and served not just to document the subject of embryological research, but also to provide it. Since embryos are hidden within the womb or the eggshell, the images on the paper made a process visible that would otherwise remain unseen. And because each embryo drawing in the table is based on a different specimen, it is clear that the resulting series of pictures enables us to see something that could never be perceived by the naked eye in nature. A different organism must be used for each stage of the process because, at the moment that the fetus is taken from the womb or egg to serve as the subject for a drawing, it dies and consequently its development stops. Plate XXX in Alexander Ecker's famous *Icones Physiologicae* from the 1850s, for example, combines views of at least four individual specimens to show the development of the human embryo; the first and youngest fetus was "a gift from surgeon and obstetrician Göz in Freiburg," the one a few weeks older was "a gift from Dr. Bloch in Emmendingen," the third "a gift from the public health official Dr. Lederle in Staufen," and so on. Combining a number of arrested developmental paths into a picture of a single individual's emergence

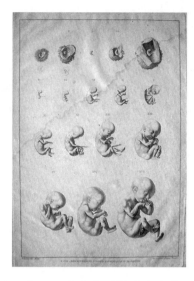

Figure 44. A plate from Soemmering's 1799 illustrated atlas

was an undertaking requiring considerable logistical effort. A laboriously collected patchwork of specimens and models is unified in the picture, becoming comprehensible at a glance. Broken down into a few stages and sequentially arranged on a sheet of paper are images of miscarriages representing nine months of human embryological development—a growth process too slow to be perceived in real time. Within a few decades, embryological research used this method to trace the development of many groups of animals: humans, frogs, rabbits, deer, fish, dogs, chickens, guinea pigs, and tortoises.[18]

In addition to Darwin's library, the naturalist's work itself attests to his deep familiarity with these pictures from the fields of anatomy, physiognomy, and embryology. Although his *Origin of Species* contained just a single illustration, a close reading shows that another picture, masked in words, found its way into the book in the form of a citation. Like the fold-out evolutionary diagram, this hidden picture was also greatly indebted to embryol-

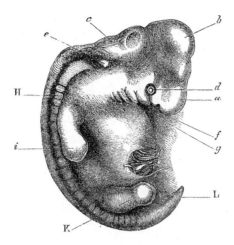

Figure 45. Human embryo in the 1871 *Descent of Man*

ogy. It appears when Darwin describes Agassiz's observation that "the geological succession of extinct forms is nearly parallel with the embryological development of existing forms." A few sentences later he refers to the embryo as "a sort of picture, preserved by nature, of the former and less modified condition of the species."[19] The words "picture, preserved by nature" do not constitute a metaphor, but rather describe the images contained in the embryological atlases housed in London's scientific libraries. Twelve years later, at the beginning of *Descent of Man*, Darwin would publish the illustration he only mentioned in 1859 (Figure 45). Excusing his earlier failure to show the embryo picture, he wrote: "As some of my readers may never have seen a drawing of an embryo, I have given one of man and another of a dog, at about the same early stage of development, carefully copied from two works of undoubted accuracy."[20] The illustrations that follow show a human embryo taken from Plate XXX of Alexander Ecker's work and a dog embryo based on a Theodor Bischoff drawing.

Without changing Ecker's model, Darwin thus gave the embryo portrait a new depth. Ecker uses the illustration to show an early stage of ontogenesis, but in Darwin's book it now has two overlapping meanings: it shows an early stage of individual development on the one hand and the beginning of familial phylogenesis on the other, as the "picture . . . of the former and less modified condition of the species."[21] The early-stage embryo thus becomes a freeze-frame in the evolution from monad to human being. Depending on whether we read the series from which the picture is taken in ontogenetic or phylogenetic terms, it encompasses either a period of nine months or—as Darwin estimated at that time—well over three hundred million years. The process of evolution, ticking with such aching slowness through millions of years that the entire history of humanity barely suffices for us to notice the gradual changes in species, is shown in the blink of an eye.

In terms of the visual tradition, Darwin brought two previously separate pictorial forms together when he—indirectly in 1859 and directly in 1871—declared the embryo to be a picture preserved by nature from the course of evolution. Embryological plates show a developmental process that unfolds in time but remains within a single species, while anatomical or physiognomical plates show representatives of several species together but do not imply a temporal development among them. Darwin, however, reads a process in time as an anatomical comparison and vice versa. The similarities among living things occurring at different times and among different genera—similarities that Cuvier, Agassiz, and Owen saw as signs of an underlying plan or principle—seemed to him to indicate recurrence. That Darwin accomplished this combinatory feat by considering the images themselves is supported by more than just his reference to the embryo as picture—the reference for which he later named Ecker's and Bischoff's pictorial atlases as sources. A note in his copy of Owen's *On the Nature of Limbs*, the book containing the plate with the archetype and the five vertebrate skeletons, also supports this view. On a blue slip of paper placed at the back of the book, Darwin wrote, "I look at Owens Archetype as more than ideal, as a real representation" (see Figure 42).[22] Like the embryo, the archetype illustration thus becomes a double portrait, simultaneously showing history and the present.

While Darwin's initial reference to the embryological plates in *Origin of Species* went largely unnoticed in the 1860s, the picture series of another author unleashed a furor. It was the work of the anatomist and physiologist Thomas Henry Huxley, whose *Evidence as to Man's Place in Nature* was the first work of evolutionary theory published after 1859. The book, which appeared in February 1863, examined the very topic Darwin had avoided: mankind's kinship with the animals. Three years earlier, at a meeting of the British Association for the Advancement of Science in Oxford, Huxley had triumphantly asserted humanity's descent from the animals—replying to an attack by Bishop Samuel Wilberforce that animal ancestors were preferable to clerical ones. Its scandalous content and pithy style made his book a best seller, one that was sold even in train stations as good reading for travelers. The public responded with a mixture of ridicule, enthusiasm, and disgust. In a letter Darwin cheered, "Hurra, the Monkey Book has come. . . . The pictures are splendid."[23] Offering a contradictory assessment was the Duke of Argyll, one of the most powerful opponents of the theory of evolution. He labeled Huxley's key image as a "grim and grotesque procession." The subject of Darwin's praise and Argyll's condemnation was Huxley's frontispiece (Figure 46), which showed an anatomical comparison of five skeletons, all from the genus of hominids. From right to left, a gibbon, orangutan, chimpanzee, gorilla, and human march past in profile. To make this comparative arrangement possible, Huxley had artist Benjamin Waterhouse adjust the scale when he drew the skeletons from the collection of the Royal College of Surgeons; each has the correct proportions, but all the images are approximately the same size.[24] In this way, the much smaller gibbon at the left edge of the picture is as big as the human, who in turn is almost as tall as the gorilla, which would tower over him in real life. The anatomical comparison of human and ape in Huxley's picture was not new in itself. For a few pence, any visitor to the British Museum could buy a photograph

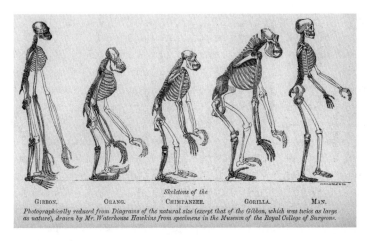

GIBBON.   ORANG.   *Skeletons of the*   GORILLA.   MAN.
                            CHIMPANZEE.

Photographically reduced from Diagrams of the natural size (except that of the Gibbon, which was twice as large as nature), drawn by Mr. Waterhouse Hawkins from specimens in the Museum of the Royal College of Surgeons.

Figure 46. Frontispiece to Huxley's 1863 *Evidence as to Man's Place in Nature*

showing the skeletons of a human and a gorilla side by side in front of a shaded background.[25] But while this juxtaposition primarily served to highlight the differences between the erect human and the seemingly coarsely built animal, Huxley's multiple comparison dislodged the human from his status as unique. As Darwin had done with the bent-, cross-, and thick-beaked finches in 1845, Huxley placed gibbon, orangutan, chimpanzee, gorilla, and human within a single image.

The human thus became one of several variants, exhibiting the same relationship to the hominid family that the finch *Geospiza parvula* could boast to the genus *geospizinae*. According to Huxley, mankind's closest relative was the gorilla. "For the skull, no less than the skeleton in general," he wrote, "the proposition holds good, that the differences between Man and the Gorilla are of smaller value than those between the Gorilla and some other Apes."[26] Showing the figures in profile furthermore gave the series of apes a dynamic quality, suggesting that the lineup showed a possible history of humanity's development. Bent slightly forward, the apes seem about to leap ahead and take on the form of the next member of the line. Linnaeus's once unassailable taxonomical hierarchy—which, despite assigning humans to the category of primates, preserved humanity's special status—was poised to become a mere historical artifact.[27] For his part, however, Huxley left the picture's interpretation open. His text does not indicate whether he intended the arrangement of skeletons to be understood as a mere anatomical comparison or a speculation about a potential process. In any case, the suggestion of temporal development contained in the picture contributed to its popularity. The apelike ancestor that stands up, sheds its fur, and becomes a human being went on to become one of evolutionary theory's most popular icons, still found today in textbooks, at exhibitions, and on posters, billboards, and stickers. Huxley provided the model for this figure, and his picture was often reprinted during the nineteenth century in books such as Haeckel's *Anthropogenie oder Entwickelungsgeschichte des Menschen* (Anthropogenie, Or the Evolution of Man), his most-published work.[28]

The amalgam of evolution and popular culture that existed from the very beginning and contributed to the theory's success is shown by the speed with which fantastic variations of the picture series appeared in the wake of Huxley's attention-grabbing frontispiece. These were often in the tradition of French caricaturists such as Grandville, who had turned humans into apes, lions, pikes, or lobsters in *Les métamorphoses du jour*, or Charles Philipon, whose famous caricature from 1831 portrayed the French king Louis-Philippe as a pear.[29] When *Evidence as to Man's Place in Nature* appeared in February 1863, the world of English caricature responded by giving birth to a veritable zoo of hybrid creatures that flapped around the Victorian public like the monsters surrounding the head of Goya's sleeping Reason. From the satirical *Punch* to the *Illustrated Times*, professors became oxen, children became monkeys, women turned into pumpkins, and parrot cages into balloons. Offshoots turned up later in children's books such as Charles Kingsley's *The Water-Babies*, illustrated in 1885 by Linley Sambourne, an employee of *Punch* who had drawn Darwin for the magazine.[30]

The most extensive body of work came from Charles Bennett, the top caricaturist for the *Illustrated Times*. From May to October 1863, issue by issue and page by page, he expanded the bestiary of evolution. The first cartoon of the series had the title *The Origin of Species, dedicated by Natural Selection to Dr. Charles Darwin*. It shows the metamorphosis of a scholar who bears a resemblance to pictures of Darwin from the period, before he adopted a full beard. The figure in the drawing first turns into an ox and then a pig. The counterclockwise movement of the circular arrangement shows the professor's sideburns becoming horns, his chin stretching into a bovine muzzle, his long formal jacket becoming a hide, and the professorial bearing deteriorating until the arms touch the floor. In the steps that follow, the ox shrinks to become a pig, and the cycle closes with a pig teetering on a bellows next to the original scholar with his book. The pig represents the professor's allegorical mirror image: the champion of the theory of evolution has become an animal himself, and his book is nothing but a means to blow hot air. The circular arrangement leaves the direction of this transformation open—the pig can just as easily become a professor as the professor a pig.

On a weekly basis, Bennett turned butlers into bulldogs, lawyers into weasels, musicians into foxes, and workers into dancing bears before the very eyes of the *Illustrated Times*'s readers. Four weeks after his initial salvo, the fifth entrant in the series, "A Monkey Trick," appeared at the end of May (Figure 47). Here, an English schoolboy turns into a monkey and takes all the surrounding objects with him, picture by picture, through a magical transformation that extends to the inorganic world as well: the parrot becomes an animal trainer and his cage turns into a balloon. This restless activity does not merely culminate in a "monkey trick" that the parrot-turned-animal-trainer performs with the erstwhile schoolboy—instead, the entire image is a trick, an example of artistic sleight of hand. The intermediary pictures that link human and ape or balloon and cage are based in reality no more than the image linking Lavater's frog and

Figure 47. 1863 caricature "A Monkey Trick"

Apollo. Caricatures show that nothing exists in the visual imagination that cannot be joined with anything else through a series
of intermediate steps. Like the bellows in the earlier caricature,
the hot-air balloon symbolizes a suspicion of speculation and
the accusation that the theory of evolution is just an audience-
pleasing thought experiment. To spell out this visual critique,
the caricature suggests that the similarity Huxley establishes
between human and ape through a series of steps may also rest
on a mere "monkey trick."

Among the fans of evolutionary theory in its caricature form
was, surprisingly, Darwin himself. His collection of pictures
referencing either him or his theory can be found at the Cambridge University archive. "Ah, has *Punch* taken me up?" he

asked a friend in 1872, admitting, "I keep all those things. Have you seen me in the *Hornet?*"[31] The caricatures apparently flattered Darwin more than they offended him, a response which supports Janet Browne's thesis that they contributed to awareness of the theory of evolution. The idea that we are descended from apes was something to laugh about in Victorian England, and such humorous reproduction extended the theory's reach beyond scientific societies or clubs and into living rooms and nurseries. The potentially insulting idea that animals are our ancestors was softened. Instead of the dominant narrative topoi of evolutionary theory—struggle and selection—the pictures revealed a different side of the theory and associated evolution with metamorphosis. As noted in this book's introduction, they thus grant Darwin's theory the flair of "a clear day in a moderate English summer," as Mandelstam stated.[32]

Despite Darwin's enthusiasm, however, the caricature touches on a sore point. Of all accusations, Darwin most feared the charge that his theory was speculative—precisely the same criticism he leveled at all previous theories of evolution. Of Lamarck's *Histoire naturelle des animaux sans vertèbres* he energetically noted: "This volume has no facts, wild metaphysical speculation." In another comment he condemned Lamarck's *Philosophie zoologique* with the words "Very poor and useless book." His reading of Chambers's *Vestiges* led him to conclude, "I will not specify any genealogies.—much too little known at present." "It is doubtful whether Lamarck has done more good by awakening subject," he summed up, "or harm by writing so much with so few facts."[33]

At a conscious remove from his predecessors, Darwin spent twenty years collecting examples and evidence before he published *Origin of Species* in 1859. The book contains a mountain of data which, to this day, his opponents have not been able to refute, yet Darwin himself admits to the "imperfection of the geological record," the fragmentary documentation of historical processes that fossils provide. "Geology assuredly does not reveal any such finely-graduated organic chain," Darwin admits, "and this, perhaps, is the most obvious and serious objection which can be urged against the theory."[34] Fossils may have revealed the order in which types of animals emerged during the earth's history: invertebrates came early, for example, and mammals evolved late. But the sparse fossil records could hardly reveal the forms that linked the new species with the old ones or how they arose from other genera, families, classes, and orders. Even the question of whether the archaeopteryx found in the Solnhofen limestone represented a link between the reptiles and birds was contentious. The caricaturists' flights of fancy aside, speculation was still unavoidable when considering these evolutionary steps.

For this reason, living species or genera often took the place of fossil specimens. Following the principle of uniformity established by Lyell in geology, phenomena in the present should be able to provide insight into processes in the past. To this end, Darwin showed four types of Galápagos finch in 1845; Huxley examined a gibbon, orangutan, chimpanzee, and gorilla in 1863, and in his famous embryo plates from 1872 Haeckel would eventually reform the embryonic development cited by Darwin into a retelling of the history of the race. Instead of fossils, contemporary species and genera were visually projected through time. The function of a series of embryological pictures to negotiate the relationship of time and change was thus carried over to comparative anatomy.[35]

The most elaborate case of this type in Darwin's work is that of the argus pheasant. With the stated purpose of reconstructing the possible stages of the argus pheasant's evolutionary history, Darwin combined the variations of the pattern he had found on the bird's plumage into a series of five pictures. They were meant to counter the claim that the pattern's perfection is proof that it is the work of the Creator. Anticipating his opponents' arguments, Darwin introduced the series with the words: "No one, I presume, will attribute the shading, which has excited the admiration of many experienced artists, to chance—to the fortuitous concourse of atoms of colouring matter. That these ornaments should have been formed through the selection of many successive variations, not one of which was originally intended to produce the ball-and-socket effect, seems as incredible, as that one of Raphael's Madonnas should have been formed by the selection of chance daubs of paint made by a long succession of young artists, not one of whom intended at first to draw the human figure."[36]

The picture series was thus meant to reconcile the seemingly irreconcilable and explain how the artistic could emerge by chance. Herschel's dismissive characterization of evolution as the "law of the higgledy-piggledy" continued to occupy Darwin: admitting that it was "a good sneer," he noted that "it made me put in the simile about Raphael's Madonna."[37] With this comparison, he chose the very artist whom the Pre-Raphaelites considered the epitome of artificiality. For this group, centered around Dante Gabriel Rossetti, William Holman Hunt, and John Everett Millais, the late Renaissance painter represented a turning point at which art rejected nature and truth for the affected artificiality of perfection. By juxtaposing the perfection of Raphael's Madonnas and the chaos of "chance daubs of paint," Darwin distilled the question of the type of picture nature presents to us. Divine authorship, perfection, and mastery stand opposed to ongoing processes, flaws, and a lack of completeness. These alternatives arose from the metaphor of art and chance, while the answer came in a the form of a picture.

GOULD'S HUMMINGBIRDS

The second strand of Darwin's net of drawings, pictures, and letters is strung along the argus pheasant and its striking plumage. This thread leads out of the elaborately constructed pictorial atlases of embryologists, physiognomists, and anatomists to a seemingly intuitive way of seeing and a simple anecdote from the Reverend William Paley. In the first pages of his 1802 book *Natural Theology*, which Paley—a Cambridge graduate like Darwin—published three years before his death, he describes a walk through a heath and two things he found there. First, he writes, he kicked up a stone. Asking himself where the stone could have come from, he reached the answer that it must have always been there on the heath. Then he found a watch. He answered the question "Where did this come from?" differently than he did the question for the stone: with its interlocking wheels and hands, the complicated mechanism of its works, its polished case, and its purpose of telling time, the watch could not have originated on the heath. The conclusion, he says, is that "the watch must have had a maker: that there must have existed, at some time, and at some place or other, an artificer or artificers who formed it for the purpose which we find it actually to answer [i.e., indicating the time]."[38] In the third chapter of his book, Paley applies this same argument to the human eye, which he compares to a telescope, and in the pages that follow he expands this line of reasoning to all of organic creation—respiratory organs, the circulatory system, joints, the structure of muscles—all the elements that in his opinion make organisms comparable to complex machines. Just as wheels, screws, and hands point to a watchmaker, machinelike organisms and the miraculous interaction of their individual parts indicate the existence of a Creator.

Paley's book, which appeared in numerous editions during the nineteenth century and which was also translated into German and French, was thus part of a tradition, reaching back to the period in the seventeenth century in which the Royal Society had been founded.[39] Since the time of Robert Boyle, the works of naturalists had been a means to assemble an inventory for a steadily growing museum of divine creation; from shells, butterflies, flowers, or lizards to birds or mammals, not a single segment of the animal kingdom, when examined closely, failed to reveal the skill and taste of its maker. In curiosity cabinets, art and nature were exhibited side by side, emphasizing the apparent intention discernible in both cases. The beauty of a peacock's feathers revealed God's existence just as the construction of the heart's chambers did. As in Paley's book, the natural object most often used to illustrate this connection was the eye. For naturalists, the finely calibrated system of light-sensitive membranes,

cells, lenses, and tendons that together made perceiving and fo-
cusing on objects possible worked with the precision of a mechanical instrument. The eye was a natural telescope. It became
the textbook example of the so-called argument from design—
proof of the existence of God derived from examining nature as
a feat of art or engineering. Thanks to the role of art and technology in the argument from design, the concept remained popular even in the industrial age; works of natural theology
remained part of the university curriculum in the nineteenth
century. In 1830, for example, as Darwin was still studying in
Cambridge, the Earl of Bridgewater, an official in the Anglican
church, offered a cash prize for a series of essays titled "On the
Power, Wisdom, and Goodness of God As Manifested in the Creation," whose authors would eventually include the mathematician Charles Babbage and the philosopher William Whewell.
And when Darwin, at the other end of the world, saw the Australian ant lion employing the same technique as its European
counterpart to catch its prey, he did not hesitate to interpret this
similarly as evidence of the "hand of the creator." Apparently
the perspective of natural theology seemed self-evident to him
as well.[40]

The extent to which this view saturated the society around
him did not ebb when Darwin returned from his journey and
felt the first twinges of doubt. In the middle of the 1840s, as he
was already working on the second draft of his theory of evolution, Whewell, a professor at Cambridge, published *Indications
of the Creator.* The book collected all the evidence suggesting
that intent and purpose were behind the forms of nature and its
objects.[41] Tellingly, the book was a response to the appearance
of Chambers's *Vestiges*—an attempt, as we have seen, to explain
the creation of animals and plants with a theory of evolution
(see Figure 24).[42]

However, the most convincing argument for natural theology
did not come from the pens of scholars; instead, nature itself
seemed to supply it. The Creator's divine handiwork could be
enjoyed by the shell collector strolling on the beach or the owner

of a pocket microscope peering into the structure of the micro-cosmos.[43] In the popular science of laymen, the argument from design fed on an enthusiasm for the wonders of nature that was part science and part edification. Strictly speaking, an observer did not even need faith to see the Creator's work in the natural world. In the logic of natural theology, humans were not moved to view nature by a belief in God; instead, observing nature led to this belief in the first place. Nature proved the Creator's existence. Just how common this view was in Victorian popular culture can be seen in the way Charles Dickens described an exhibition in 1851. Almost fifteen years had passed since Darwin's journey when, parallel to the first World Exhibition at London's Crystal Palace, a pavilion with five thousand stuffed hummingbirds opened in Regent's Park. In an article titled "The Tresses of the Day Star," published in his magazine *Household Words*, Dickens compared nature and art in the form of the work of the Creator in Regent's Park and that of earthly engineers in the Crystal Palace. "Study the useful and ornamental inventions of the civilised world," he told his readers, referring to the World Exhibition, "but study, too, the work of the Divine hand in these little birds."[44]

More than seventy-five thousand paying visitors, including Queen Victoria and Prince Albert accompanying the Count and Countess of Saxony-Coburg and Count Ernst von Württemberg, heeded this call and squeezed along the pavilion's twenty-four showcases containing insect-sized birds positioned against a background of branches and leaves. With the help of wires, the tiny creatures hovered before fuchsia or bromeliad blossoms to make them appear as lifelike as possible, and Dickens wrote that he thought he could hear the buzzing of their beating wings. According to Dickens, these works of nature in preserved form were in no way overshadowed by works of art. A collector who could own such a hummingbird, he wrote, would be as deserving of admiration and envy as the owner of a Correggio.[45] In other words, a hummingbird was as valuable as a painting by one of the most famous painters of the Renaissance.

The nearly complete representation of all the world's hummingbird species allowed England to celebrate not just the oneness of art and nature, but its own status as an all-embracing nation at the heart of which the rest of the world could be found *en miniature*. Like gold or a jewel, a preserved animal brought back from America symbolized a seagoing state's lust for adventure, a commercial power's pride in ownership, and the exoticism of foreign lands where nature possessed luxuriant and mysterious beauty. English ladies soon followed the fashion of decorating their clothing and hats with hummingbird feathers or even entire birds, while the gentlemen of the royal cavalry sported a band of ostrich feathers on their helmets. In the nineteenth century, more hummingbirds and other exotic birds could be seen in London than anywhere else on earth.[46]

Only hints exist as to whether Darwin visited the hummingbird exhibition. In order to bring the manuscript of his *Monograph of the Sub-class Cirripedia*—a treatise on barnacles that earned him the Royal Medal—to his publisher John Ray, Darwin traveled to London with a his family for a week. His wife's diary for 1851 describes their visit to the World Exhibition, a diorama in Regent Street showing the route between London and Calcutta, and the Zoological Gardens, where the hummingbird pavilions stood. During this week, or shortly thereafter, the exhibition led to a reunion with an old acquaintance from earlier years: John Gould. Contact between the two men broke off after their work together on the *Beagle* collection and Gould's abrupt departure for Australia. Now the hummingbirds had put Gould back in the spotlight. He was the owner of the enchanting natural specimens on display in the Zoological Society pavilions in Regent's Park, which would also serve as the basis for a major book project. The gardener's-son-turned-institutional-head had now become a respected ornithological author and collector who would leave behind thirty-eight thousand preserved birds at his death, including more than five thousand hummingbirds alone.[47] In his article Dickens honored the exhibition's creator no less than the exhibition itself, and like Giorgio Vasari, who once described the life of Correggio, he celebrated the rise of a young man of simple means who rose to be a master of natural history. But for Darwin, Gould's once positive influence became an opposing force. The man whose identification of the Galápagos species inspired Darwin's first musings about evolutionary theory soon used the tiny birds to fuel the arguments of evolution's foes: the luxurious publication *Monograph of Trochilidae* that followed the Regent's Park exhibition came to serve as evidence against the theory of evolution. The five volumes, issued in several installments between 1849 and 1861, doubled the extent to which the stuffed birds were seen as art objects. In his introduction, which was printed separately and bound in red cardboard with gold lettering, Gould quoted all naturalists who had ever compared hummingbirds to works of art, precious stones, or machines. One was the American ornithologist and bird artist John James Audubon, who questioned whether any person "on observing this glittering fragment of the rainbow, would not pause, admire, and turn his mind with reverence towards the Almighty Creator, the wonders of whose hand we at every step discover, and of whose sublime conceptions we everywhere observe the manifestations in this admirable system of creation?" Also quoted was Georges-Louis Leclerc, Comte de Buffon, the great French naturalist of the eighteenth century, who claimed that the hummingbirds were, of all creatures, "the most elegant in form and the most brilliant in colour," putting all human inventions to shame.[48] It was no accident that beauty, luxury, and elegance were not just qualities of the hummingbirds, but also the very things that wealthy collectors valued—the type of buyers who were Gould's target audience. Not only did his ornithological plates make the birds in the lithographs look like luxury items but, thanks to a technique Gould patented in 1851, they were luxury items in themselves. When his volumes were shown at the World Exhibition in the Crystal Palace in the "Fine Arts" category, the catalog entry for exhibitor 247 read: "Gould, J. 20 Broad Street, Golden Square— Inventor. A new mode of representing the luminous and metallic colouring of the Trochilidae, or humming birds. The effect is produced by a combination of transparent oil and varnish colours over pure leaf gold laid upon paper, prepared for the purpose."[49] The costly gold leaf technique lent new authority to the comparison of art and nature; the manufactured item was as luxurious as the nature it represented. It also represented the culmination of a series of printing innovations to emerge during the course of ornithological history: from Bewick's woodcuts in the *History of British Birds* of 1797–1804 to Audubon's double-elephant format for the *Birds of America* of 1827–1838 to Swainson's *Zoological Illustrations* of 1820–1832, an affordable book showing almost two hundred animals, mostly birds, in color long before lithographic techniques became the standard way to produce full-color book illustrations.[50]

Using the new technique, Gould depicted four hundred and thirty hummingbird species within the four hundred pages of *Monograph of Trochilidae*, presenting two or more birds in each illustration. The hummingbirds swarmed onto the pages from all directions, sticking their bills into the calyx of a passion-flower, landing on a heliconia, disappearing entirely into an orchid blossom, brushing the bells of a hyacinth, or perching as chicks in a nest the size of a demitasse. Page after page, they fluttered around exotic new plants (Figure 48 [Color Plate 12]). Against the bird's tiny bodies, a cactus looks like a spike-covered planet, while a blossoming water lily looms from the pale-blue water like a giant artichoke. With the exception of a few flowers, the leaves, twigs, and blades of grass that form the humming-bird's playground were rendered by Gabriel Bayfield, Gould's colorist of many years, in cool pastel tones. The drawings and print materials were the work of Henry Constantine Richter and Edwin Charles Prince, who had taken the deceased Eliza-beth Gould's place in her husband's operation in 1841.[51] The real eye-catcher, however, was the hummingbirds' plumage. The birds all sported the patented coating, on either individual spots or their entire bodies. As a result, the cheeks and cap of *Panop-lites matthewsi* shimmered yellow-green, the ruby-red throat of

CHLOROSTILBON PORTMANNI.

Figure 48. Hummingbirds from Gould's *Monograph of Trochilidae*, 1849–1861 (With the kind permission of the Berlin Natural History Museum) (Color Plate 12)

Aphantochroa golaris seemed to glitter in the sun, and the velvety black of *Loddigesia mirabilis*'s long tail feathers appeared to absorb the light. The addition of gold created the glossy effect, while the oil was responsible for the shimmering surfaces. Since the reflecting colors made it difficult to create depth with gradations in color or shading, the birds appear remarkably flat. Looking at the pictures is thus rather like peering through a stereoscope: instead of appearing three-dimensional, the creatures and plants seem like cutouts, and the contrast between the soft pastels of the background and the shiny metallic birds produces an illusion of space between them. In the course of the five volumes, the glazing technique was constantly improved and refined, and the birds' colorful throats, undersides, backs, crowns, and caps sparkle more and more. The painted birds grow further and further removed from their real-life models and take on an independent existence that could hardly be more artificial. By the fourth volume, the birds buzz across the pages like tin toys, tracing stiff loops or stalling in the air with their stiff, glittering wings. The robotic touch with which they are portrayed corresponds to Gould's observation that the buzzing of their wings reminded him of the sound of "a piece of machinery acted upon by a powerful spring."[52]

Alongside Alexander von Humboldt, the subscribers to *Monograph of Trochilidae* included aristocrats such as the Prussian crown princess, to whom the book was dedicated; Queen Victoria; the royal houses of Saxony, Hanover, Denmark, Belgium, and Portugal; and the princess of Wied. Libraries around the world ordered the work: from Oxford to Melbourne and from Paris to Saint Petersburg, as well as libraries in Guatemala, Vienna, Berlin, New York, and British Guyana.[53] Gould dubbed one hummingbird *Eugenia imperatrix* in honor of the wife of Napoleon III. By the end, the names of virtually all the crowned heads of Europe graced the list of subscribers.[54] However, the work guaranteed that its author would enjoy not just social success, but scientific recognition as well. As late as the eighteenth century, when the twelfth edition of Linnaeus's *Systema Naturae* appeared, zoologists were aware of just eighteen species of hummingbird. Thanks to Gould, that number had grown to more than four hundred.[55] His service to science prompted numerous scientific societies and associations to grant him membership, including the Royal Society in London and the German Ornithological Society.

Although Darwin was not among the work's subscribers, he owned a copy of the introduction to the first volume of *Monograph of Trochilidae*, including a comprehensive list of species, which Gould sent to him in 1861 and which was published the same year in the journal *Athenaeum*. Darwin found himself listed alongside Prince Maximilian of Wied, Gosse, Wallace, Bates, and Salvin among the almost two dozen researchers thanked by the author for their contributions to the study of these "gems of creation."[56] Darwin's modest part in this effort was the fact that he had brought three hummingbird species back from his voyage and described them in *Zoology of the Voyage of the H.M.S. Beagle*—without, however, providing illustrations. As he read the text, it may have surprised him that Gould mentioned his own work and characterized the hummingbirds as gems in the same sentence. The comparison, after all, rested on the perspective of natural theology, which viewed nature as a work of art—an idea that *Origin of Species* opposed. Despite Gould's claim that his account was "without bias, one way or the other, as to the question of the Origin of Species,"[57] two passages in particular in which Gould referred directly to the theory of evolution may have taken Darwin aback. These two statements, if accurate, would have been blows to the very heart of evolutionary theory. First, Gould stressed his belief that the hummingbird's coloration was purely decorative and that it did not provide the birds with any advantage in their environment. Furthermore, he considered the colors of their plumage, like all the birds' other characteristics, to be constant, not subject to deviation or change. Gould claimed that he failed to encounter a single species variant in his twelve years of uninterrupted work on the subject (the volumes appeared from 1849 to 1861).

According to one colleague, the ambiguity with which Gould argued against Darwin while claiming not to was a calculated entrepreneurial stance. "It is most amusing to see how anxious he is to avoid committing himself about Darwin's theory," wrote the ornithologist Alfred Newton, who also corresponded with Darwin. "Of course, he does not care a rap whether it is true or not—but he is dreadfully afraid that by prematurely espousing it he might lose some subscribers, though he acknowledged to me the other day he thought it would be generally accepted before long."[58]

Gould's protests did not convince Darwin. Inside the red cardstock cover of his copy, he asked himself if citing "all these cases of doubt" would be worthwhile. He was referring to cases in which even Gould himself was not sure whether certain specimens represented new species. These examples led Darwin to conclude that these hard-to-characterize birds were varieties and that Gould had thus witnessed the very phenomenon he claimed never to have seen during his research. With this idea in mind, he politely thanked the author with a letter in which he remarked, "I am extremely much obliged to you for your present of the Trochilidae, of which I have read every word . . . from your pleasant Introduction to the end. . . . One of the points which has interested me most (which you will not approve of) is the number of 'races' or doubtful species. I think I shall extract all these cases; as it will show those persons who are quite ignorant of Nat History, that the determination of Species is not a simple affair."[59] In this way, Darwin derived the same lesson from Gould's classification in 1861 that he had twenty-four years earlier in 1837, when Gould identified the Galápagos finches.

ARGYLL'S ARGUS PHEASANT

In 1867 Darwin's plans to analyze Gould's work and distill a cautionary lesson about the pitfalls of taxonomy and the range of variation in nature were interrupted by the appearance of yet another publication.[60] The author was George Douglas Campbell, the eighth Duke of Argyll, and the book was *The Reign of Law*. As we have seen, the duke had condemned Huxley's series of hominids on the frontispiece of *Evidence for the Place of Man* as a "grim and grotesque procession"; now he dedicated an entire book, excerpts of which he had already published in journals, to criticizing the theory of evolution.[61] These attacks forced Darwin to again take up pen and paper and follow the finch picture and the diagram with another systematic view of evolution. Shortly after the publication of *Origin of Species*, the duke— who would later serve as a pallbearer at Darwin's funeral in Westminster Abbey—voiced his criticism for the first time. These objections came regularly and with increasing intensity whenever a new work of evolutionary theory by Darwin or one of evolution's other proponents appeared.[62] He made a name for himself on the political stage in a number of offices, including secretary for the Indian colonies. *The Reign of Law*, published in 1867, marks his debut as an author. The book recapitulates all previous objections to evolution and adds many new ones, including Gould's luxurious hummingbird books, which, according Argyll, argue against the theory. His trenchant and elegant style impressed even his opponents. "The Duke's book strikes me as very well written, very interesting, honest & clever & very arrogant," Darwin remarked in a letter. "Among the various criticisms that have appeared on Mr. Darwin's celebrated 'Origin of Species,'" conceded Wallace in a discussion in the *Quarterly Journal of Science*, "there is, perhaps, none that will appeal to so large a number of well educated and intelligent persons as that contained in the Duke of Argyll's *Reign of Law*."[63]

Argyll unerringly aimed his criticism at the gaps Darwin had left in his text and images. On the one hand, he criticized the word "origin" in the title as misleading, since the book failed to address the origin of species at the beginning of evolution and examined only the process of variation and selection that affects existing species during the course of natural history.[64] A glance at the diagram, which unlike Darwin's drafts includes no point of origin, confirms Argyll's claim. On the other hand, the duke raises the criticism that chance and natural selection are insufficient to explain many phenomena in the animal kingdom. He claimed that the colorful plumage of the male humming-bird, as presented in Gould's voluptuous pictures, makes little sense in the light of evolutionary theory, saying that "[a] crest of topaz is no better in the struggle for existence than a crest of sapphire."[65] In Argyll's eyes, Gould's gold-leaf hummingbirds became a counterrevolutionary manifest in painted form. At the same time, he repeated the author's comment that the birds' plumage is decorative only. He referred to *Monograph of Troch-ilidae* at a number of points and added another bird with an ap-pearance that seemed equally breathtaking and pointless: the argus pheasant.

This creature, named for the hundred-eyed guardian of a beloved of Zeus, carries an ornament on its wings that Argyll presented as a living rebuttal to evolution (Figure 49 [Color Plate 13]). One passage reads:

> There is one instance in Nature (and, as far as I know, the only one) in which ornament takes the form of pictorial representa-tion. The secondary feathers in the wing of the Argus Pheasant are developed into long plumes, which the bird can erect and spread out like a fan, as a Peacock spreads his train. These feath-ers are decorated with a series of conspicuous spots or "eyes," which are so coloured as to imitate the effect of balls. The shad-ows and the 'high lights' are placed exactly where an artist would place them in order to represent a sphere. [. . .] The "eyes" of the Argus Pheasant are like the "ball and socket" ornament, which is common in the decorations of human art.[66]

In the same year, John George Wood's popular *Illustrated Natu-ral History* invoked art when it mentioned the argus pheasant in conjunction with a dizzying array of shades. "Jetty black, deep rich brown, orange, fawn, olive and white are so justly and boldly arranged," the text claims, "as to form admirable studies for the artist, and totally to baffle description."[67] For both Wood and Argyll, the argus pheasant was not just a divine machine like all organisms; it was also marked by a divine artwork—a

Figure 49. A stuffed argus pheasant in the form of a fireplace screen in a photo belonging to Darwin (Darwin Archive of the Down House Museum, English Heritage) (Color Plate 13)

trompe l'oeil figure resembling the ball-and-socket ornaments often used to decorate the capitals of columns. Many examples of protective patterns exist in the animal kingdom mimicking the appearance of the environment or of other creatures. But only that of the argus pheasant, as the duke correctly suspected, imitates a three-dimensional object, its seeming depth copied by nature as convincingly as a master artist reproduces objects in a still life.[68]

It seemed obvious to Argyll that evolution could not explain such an ingenious ornament. How could such a structure occur by chance? What advantage could a trompe l'oeil offer for an argus pheasant's survival? The natural picture on its feathers, Argyll argued, was the handwriting of the Creator, not a trace of evolution. The counterargument to the theory of evolution that Argyll presents in *The Reign of Law* largely resembles Owen's archetype theory. Argyll did not dispute the idea that nature has a history. He also acknowledged that the earth's geological layers show a succession from simple organisms to more complex ones. However, he denied that these changes were due to natural selection and instead advocated what he called "creation by law"—in other words, a methodical series of acts of creation directed by God. Remaining unresolved was the question of whether existing creatures possessed an inner developmental drive or if genuinely new acts of creation occurred. However, the duke insisted that this process was a teleological one in which organisms developed toward a particular goal. He considered the evolution of the horse and cow as examples, citing the fossil histories Owen had reconstructed as proof of the animals' increasing refinement.[69]

The selection of natural objects found in *The Reign of Law* thus seemed irreconcilable with Darwin's examples. The structural similarities among a whale's fin, a human hand, and a bat's wing or rudimentary, nonfunctioning organs such as the appendix or the nipples of male mammals were the kinds of evidence Darwin painstakingly compiled in *Origin of Species* as proof of evolution. Argyll, in contrast, focuses his gaze solely on the beauty and complexity of the organic world. Compared with Darwin's bourgeois mercantile view, in which animals and plants compete, struggle, and either prevail or founder, Argyll displays an aristocratic sensibility—a taste schooled on palaces, landscaped gardens, grand tours, painting, and architecture.[70] Judging from their descriptions of nature, the books seem to deal with two different worlds. In not a single case did one man see what the other described. While Darwin discovered the back and forth of nature in flux, a chaos of difference, chance, and selection, the duke passed through an enormous open-air museum in which the masterpieces of creation were displayed like civilization's achievements in the halls of the World Exhibition. In the filigree wings of the praying mantis he saw the windows of a Gothic cathedral, the plumage of the hummingbird glittered in his eyes like jewels, and he praised the design of the tall lichens growing on ocean cliffs. For him, the beating wings of swallows and falcons were machines, and the organ that gives the electric eel its name is, in his words, a "battery," albeit one "greatly exceeding in the beauty and compactness of its structure, the batteries whereby Man has now learnt to make the laws of Electricity subservient to his will."[71] But what the argus pheasant displayed on its wings outshone all these wonders (see Figure 40). Like Paley's stroller on the heath or Dickens's exhibition visitor, Argyll's reader need not invest any belief in God before recognizing His work in the argus pheasant; instead, it was the argus pheasant itself that instilled faith. In Paley's analogy of machines, nature testified to the existence of a divine engineer, while in Gould's, Dickens's, and Argyll's artwork analogy it proved the necessity of a divine artist.

Darwin knew from his own experience during the *Beagle* voyage just how strongly the argument from design could weigh against evolutionary theory, and he addressed this point in *Origin of Species*. Two days before the book appeared he admitted in a letter that he once knew William Paley's *Natural Theology* almost by heart.[72] In a reversal of Paley's claim that the eye alone was sufficient to prove the existence of God, he wrote in 1859 that he would consider his theory dashed by a single product of nature that could not be explained by evolution.[73] A few months after *Origin of Species* was published, he confessed which object he had in mind in a letter to the American botanist Asa Gray. "It is curious that I remember well time when the thought of the eye made me cold all over," he confessed, "but I have got over this stage of the complaint, & now small trifling particulars of structure often make me very uncomfortable. The sight of the peacock's tail, whenever I gaze it, makes me sick!"[74]

From his earlier reading of *Natural Theology*, Darwin may have remembered the gist of Paley's words on the role chance could play in the natural world, and perhaps he even knew them by heart: "What does chance ever do for us? In the human body, for instance, chance, i.e. the operation of causes without design, it may produce a wen, a wart, a mole, a pimple, but never an eye."[75] Paley may have considered chance as incapable of producing nothing more than warts, pimples, or moles, but Darwin assigned it a key role—along with selection—in the creation of new species during the course of evolutionary history. This central position was valid even in the case of the nausea-inducing ornamental peacock's eye, not to mention the far more daunting one of the argus pheasant.

Argyll's hummingbird-and-argus-pheasant-based objections to the theory of evolution rested primarily on the striking plumage of the males. Shortly after the *Reign of Law* appeared in 1867, Darwin set out for the British Museum, where he subjected the specimens kept there to a thorough examination. The fantastic plumage of the bird, native to the Malaysian tropics, had made it a popular trophy of the English colonial rulers, so the museum had five specimens, including one Wallace had collected on Borneo in 1862. Instead of the Creator's handwriting, though, Darwin was looking for traces of evolution.

Darwin embarked on his search with a pencil and paper, the natural image on the argus pheasant's feathers before his eyes and taking shape under his hand (see Figures 37–39). Since the argus pheasant itself generated the natural image that graced its wings, Darwin's drawings actually represent a directed process of reproduction that created the picture a second time. The act of drawing thus led to a remarkable reconfiguration of the objects figuring in the debate. The positions of eye, eye ornament, nature, and artist began to circulate: Darwin assumed nature's place as artist redrawing the naturally occurring picture found on the argus pheasant's feathers. And the researcher himself was a product of nature tracing the evolution of a fellow product: according to Darwin's theory, the drawing hand and seeing eye are formed by chance and selection just as the argus pheasant, its ornament, or the hen watching the male perform his mating dance are formed. Eye, eye ornament, and researcher entered into a new relationship at Darwin's desk—one that would lead to the picture series illustrating the theory of evolution for *Descent of Man* in 1871.

The documents in the Darwin archive in Cambridge—a loose collection of notes and sketches—make it possible to reconstruct this encounter even though the pages do not indicate where Darwin prepared these drawings. They could have been made during his visits to the British Museum, eye to eye with the collection's specimens, against a background of busy activity of the specialists, guards, and staff—people he turned to with numerous questions during his research. But they could just as well have been drawn in his study, in the seclusion of his country house in County Kent, where he had several feathers sent for perusal.[76] At home in Downe, he drew on information from correspondents from Jena to Java, whose replies to his inquiries arrived daily (Figure 50 [Color Plate 14]). While the pages do not reveal the location where the drawings were prepared, Darwin's gaze is inscribed within them. He now sought in the wings of the argus pheasant what he had indicated in the diagram with branching lines: chance variation. In the course of this effort he began to see something completely different from what the duke had concluded. The indivisible form of a natural artwork that Argyll had perceived broke down under Darwin's gaze into elements, geometrical forms, and spatial relationships.

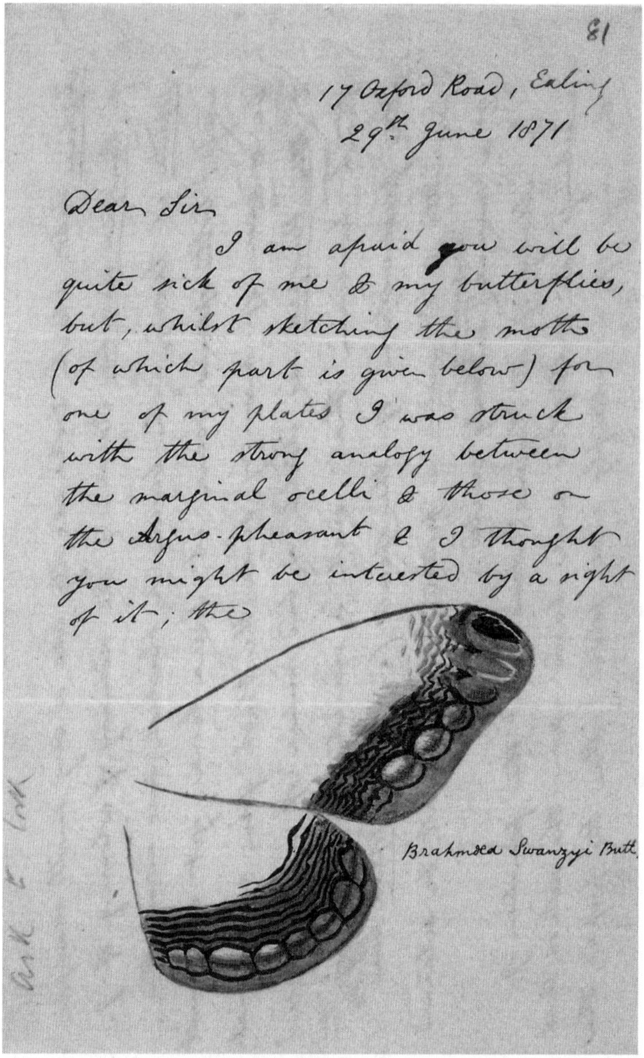

81

17 Oxford Road, Ealing
29th June 1871

Dear Sir

I am afraid you will be quite sick of me & my butterflies, but, whilst sketching the moth (of which part is given below) for one of my plates I was struck with the strong analogy between the marginal ocelli & those on the Argus-pheasant & I thought you might be interested by a sight of it; the

Brahmoea Swanzyi Butt.

Figure 50. Color sketch of an ocellus on the wings of a moth in a letter sent to Darwin (Darwin Archive [DAR 89.81], with the permission of the Cambridge University Library Syndicate) (Color Plate 14)

With each stroke, nature's figurative images became abstract compositions on the light-blue paper. Seeming representation gave way to stark silhouettes; the trompe l'oeil warmly shimmering in earth tones became a black-and-white composition of curves, points, lines, rings, and ellipses. The blots of blue ink and the rough lead pencil were anathema to everything Argyll had in mind when he compared the ornament on the pheasant's wings to an artwork and praised its perfection. A second step followed this act of drawing. Spread across six pages, the sketches appear as woodcuts by George Henry Ford in the second section of *Descent of Man* (see Figure 41).[77] What has survived in the archive as a collection of loose sheets is now a series showing the step-by-step evolution of the ornament from blotches into an ocellus. Darwin took the approach of temporalizing contemporaneous natural objects he had learned from applying embryological picture series to questions of comparative anatomy and applied this to the pattern on the bird's feathers. Picture by picture, the pages of the 1871 book decipher the process that remained encapsulated in lines in Darwin's 1859 diagram. A single natural object now appears in close-up across time, spread out in its evolutionary stages.

However, to see Darwin's picture series purely in the tradition of this image type is to overlook the formal uniqueness that breaks with its predecessors. Unlike the anatomical, physiognomical, or embryological picture series that came before, Darwin's images are spread over a number of pages—so many that readers often fail to notice their relationship at first glance. Darwin made use of the tradition's abilities to speed up time and resolve differences into a chain of similarities. But he did not adopt the structure of these earlier works. Instead, Darwin made an idiosyncratic choice when positioning the images within the book: he began with the most elaborate of the pheasant's ornaments, then showed the simplest, and then showed the stages back to his starting point with the remaining three drawings (Figure 51). Instead of following a sequence in time, the pictures change temporal direction, starting with the last stage of the evolutionary process before jumping back to the first and then building from this point toward the end. Furthermore, the pictures are not distributed consistently throughout the pages. The explanatory text provides a full-page interruption of the sequence between the third and fourth picture, and the illustrations shift from top to bottom and grow larger and smaller from one page to the next. The resulting sequence emerges in starts. Broken up by irregular blocks of text and changing direction in time, evolution appears to be hesitantly finding its way; without the teleological development shown in embryological atlases, Darwin's picture series seems turbulent, as if something unexpected could happen with each turn of the page. Darwin never tired of insisting that his illustrations in *Descent of Man* showed only how the pheasant's plumage might have evolved—no one could know for sure what had really occurred.[78]

But even as a hypothesis, the illustrations provided a counterargument to the claim that the ornament on the bird's wings could have come into being only through a conscious act of creation. From one picture to the next, a line splinters, two blotches merge, a dot stretches into an ellipse. The simple, elongated spots of the first picture change shape in the second as their pointed ends curve upward. A light shading appears in the space within these curves, which swirl together in the third picture to form a ring. To the right, emerging dotted lines gradually fuse into meandering stripes. The frame of the outline drawing and index that corsets each picture draws the observer's attention to the structure of the ocellus, and the argus pheasant's eyelike patterns remain fractured into their components at every step. When, in the fifth picture, the gray tones within the ring form the reflective surface of the ocellus (see Figure 51), nothing requiring divine intervention has occurred. The seemingly immaculate ornament that ends the series is composed of points, lines, dots, and curves and has taken shape in the course of natural changes just as a coral reef or a mountain range would.

While the reader could follow how the chance variations combined little by little to form a three-dimensional image—in

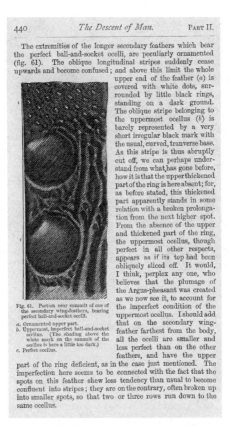

440   *The Descent of Man.*   Part II.

The extremities of the longer secondary feathers which bear the perfect ball-and-socket ocelli, are peculiarly ornamented (fig. 61). The oblique longitudinal stripes suddenly cease upwards and become confused; and above this limit the whole upper end of the feather (*a*) is covered with white dots, surrounded by little black rings, standing on a dark ground. The oblique stripe belonging to the uppermost ocellus (*b*) is barely represented by a very short irregular black mark with the usual, curved, tranverse base. As this stripe is thus abruptly cut off, we can perhaps understand from what has gone before, how it is that the upper thickened part of the ring is here absent; for, as before stated, this thickened part apparently stands in some relation with a broken prolongation from the next higher spot. From the absence of the upper and thickened part of the ring, the uppermost ocellus, though perfect in all other respects, appears as if its top had been obliquely sliced off. It would, I think, perplex any one, who believes that the plumage of the Argus-pheasant was created as we now see it, to account for the imperfect condition of the uppermost ocellus. I should add that on the secondary wing-feather farthest from the body, all the ocelli are smaller and less perfect than on the other feathers, and have the upper part of the ring deficient, as in the case just mentioned. The imperfection here seems to be connected with the fact that the spots on this feather shew less tendency than usual to become confluent into stripes; they are on the contrary, often broken up into smaller spots, so that two or three rows run down to the same ocellus.

Fig. 61. Portion near summit of one of the secondary wing-feathers, bearing perfect ball-and-socket ocelli.

a. Ornamented upper part.
b. Uppermost, imperfect ball-and-socket ocellus. (The shading above the white mark on the summit of the ocellus is here a little too dark.)
c. Perfect ocellus.

Figure 51. The final stage of the picture series on the argus pheasant in *Descent of Man*

other words, how "daubs of paint" could become a masterpiece—the pictures also implied the existence of an entity that drove the evolutionary process. The observer has the position that, in nature, is filled by the hen. Darwin showed her as well, in a picture taken from an issue of the popular natural history weekly *The Field, the Farm and the Garden: The Country Gentleman's Newspaper* from March 1874 (Figure 52). This woodcut, which filled half a page in the large-format newspaper, shows an argus

Figure 52. The argus pheasant's mating dance in an 1874 issue of *The Field*. Darwin included this illustration in the second edition of *Descent of Man* the same year

pheasant bowing in the mating dance as a hen walks past and turns her head to look at his fanned wings. The scene depicted was witnessed by animal artist Thomas William Wood in Regent's Park, where for three years two pairs of argus pheasants had been kept.

The picture clearly shows that the ornament's rightful observer is not, as Argyll suggests, a human with the ability to recognize artworks in nature, but rather a pursued and wooed hen. The force of chance in nature provides variation, but it is the hen—who chooses to mate with the male sporting the most impressive feathers—who makes the selection. The beauty of the ornaments was no more intended to impress humans than, as Darwin noted, a country fair would be planned to entertain extraterrestrials. "With respect to female birds feeling a preference for particular males, we must bear in mind that we can judge the choice being exerted, only by analogy," Darwin wrote. "If an inhabitant of another planet were to behold a number of young rustics at a fair courting a pretty girl, and quarrelling about her like birds at one of their places of assemblage, he would, by the eagerness of the wooers to please her and to display their finery, infer that she had the power of choice. Now with birds, the evidence stands thus: they have acute powers of observation, and they seem to have some taste for the beautiful both in colour and sound." According to Darwin, the animals bred the beauty they possessed on their own as a result of their "power of choice." His views thus contradict those of Argyll and all his naturalist predecessors who saw beauty as a sign of a Creator, as well as attempts within evolutionary theory today to interpret beauty as a sign of health or strength. His remarkable confidence in the abilities of animals makes him almost unique in the history of science. If humans and animals were related, then it seemed reasonable to him that animals could have an aesthetic sense.[79]

Eleven of the twenty-one chapters in *Descent of Man* examine this type of "sexual" selection, which Darwin now introduced in addition to "natural section." Four chapters concerned the "secondary sexual characteristics of birds," of which the ornament the male argus pheasant presents in the mating dance received the most attention.[80] Darwin's publisher, John Murray, did not share the enthusiasm with which his famous author increasingly explained animal characteristics as due to sexual selection. He found the term too risqué to include in the book title, as Darwin originally wanted, banning it to the subtitle instead, and requested that a particular passage on the subject be rewritten: "I wd. suggest that it might be toned down—as well as any other sentences liable to the imputation of indelicacy if there be any."[81] Murray's objections partly explain the aforementioned gap between the book's title and the images it contains. If Darwin had included "sexual selection" in the title along with "descent of man," the large number of illustrations of birds would have seemed less remarkable. However, this observation still fails to account for the absence of any pictures of human beings

whatsoever. After all, the last three chapters examine sexual selection in relation to humans, applying some of the comments
made previously about birds to human behavior. Referring to
different species of hummingbirds, Darwin had noted that "the
taste, as we see, of the several species certainly differs." Later,
discussing the preference or dislike of various peoples for
beards, he writes, "We thus see how widely the different races of
man differ in their taste for the beautiful."[82] The bower bird's
richly decorated nest, the umbrellabird's crest, the orange wat-
tles of the grouse, or the ornament of the argus pheasant—
examples Darwin provides to demonstrate how aesthetic taste
among birds varies—are all shown in pictures. But the large
beards of Anglo-Saxons, Middle Easterners, or Fiji Islanders,
mentioned to illustrate how the idea of beauty differs among
humans, appear no more than the beardless faces of the men of
Tonga and Samoa, where Darwin claims that facial hair is considered disgusting.[83] Darwin apparently made a conscious decision not to show human beings in a book called *Descent of Man*,
despite its title. The implications of this decision are examined
in the next chapter in relation to the photographs of people included in his next book, *Expression of Emotions*.

THE AESTHETICS OF CHANCE

The claim that the argus pheasant's appeal for a human observer is just a by-product of the real purpose of the bird's appearance had further consequences. With the concept of "sexual
selection," Darwin had introduced a view of nature that suggested that a large share of the characteristics of animals were
due to the preferences of these animals' mates. The male argus
pheasant was beautiful because its appearance pleased the hen.
According to this argument, hens had chosen the most impressive cock for thousands of generations until, finally, the heirs of
more and more beautifully patterned pheasants came to sport
the oscillating wings that turn the heads of the hens living
in the Malaysian tropics today. The hen's perception was thus

selective but, as Darwin added in a decisive twist, not necessarily complete. "Nor need it be supposed that the female studies each stripe or spot of colour," he warned, "she is probably struck only by the general effect."[84]

Beauty in the eye of the animal beholder must not be comparable to that of Raphael's Madonna. A thorough examination of the ornament would reveal its imperfections. Considered in this light, the final image in the picture series finally dissolves the antithesis between spot and complete trompe l'oeil. The picture series had reduced this opposition to a progression of step-by-step similarities so that the final image now revealed the flaw in the seeming perfection. In his description of the book's Figure 61, the last image in the series, Darwin remarks: "The uppermost ocellus, though perfect in all other respects, appears as if its top had been obliquely sliced off." The illustration indeed reveals the ornament to be slightly askew: instead of closing in a soft curve, the ring makes an abrupt diagonal. Furthermore, according to Darwin, the reflective spots in the trompe l'oeil are not placed consistently. These observations lead him to a far-reaching conclusion. In response to Argyll's claim that the perfection of natural objects can be explained only by the existence of a Creator, Darwin counters that this perfection itself is a myth.

The picture provides the proof of the eye-shaped ornament's imperfections; to prove the imperfection of the human eye—which until then had been considered the epitome of natural perfection—Darwin invokes Hermann von Helmholtz, "the highest authority in Europe on the subject." He quotes from the famous German physicist and physiologist's essay "The Optical Apparatus of the Eye," in which Helmholtz makes the damning judgment "that anyone to whom an optician had sold such a negligently built instrument would feel fully justified in returning it."[85]

Helmholtz claimed that the eye misperceives colors and shapes in a number of ways. Both eye-shaped ornament and eye—the two objects that, Darwin confessed in 1860 to Asa Gray, caused him to grow queasy—were given new significance in

1871 in *Descent of Man:* they were united by imperfection. This view turned on their heads both Paley's comparison of the eye to an instrument and that of the pheasant's ornament to a work of art. In terms of Darwin's argument, flaws were to the theory of evolution as flawlessness was to a Creator. No one would consider God to be the author of an imperfect object. Perfection revealed the hand of God; error betrayed the workings of nature.

As in the case of the evolutionary diagram, the relationship here between image and text broke down. In his commentary that surrounded and sometimes interrupted the picture series, Darwin tried in vain to capture the illustrations in words. He characterized the variation between two pictures, for example, in contradictory terms as "an absolutely insensible gradation." With apologies to his reader, he labeled the resulting design "for the want of a better term, an 'elliptic ornament.'" Speaking of the next picture, he was forced to admit that "between one of the elliptic ornaments and a perfect ball-and-socket ocellus, the gradation is so perfect that it is scarcely possible to decide when the latter term ought to be used."[86] This wrestling with terminology marks an inversion of the usual relationship between word and image. It is not the image that complements the text, but rather the text that complements the image. The granting of names and descriptions simply illustrates the fine points of the evolutionary process that the pictures document. The words point to the picture, where the real argument is located. Darwin's central evidence for evolution in *Origin of Species* was that the myriad variations among species, subspecies, varieties, and individuals surpassed the capacity of systems of classification to organize them. Just as a nature in constant flux evades the labels and order of a classificatory system, the process of evolution cannot be captured within a terminological grid. The stages in the evolution of the bird ornament are too subtle to be named— it is possible only to show them.

With the picture series on the argus pheasant in 1871's *Descent of Man,* Darwin thus developed the third systematic view of evolution after the finch illustration and the evolutionary diagram. In contrast to the other illustrations in the same book, such as a picture of a somewhat pointed human ear whose shape, according to Darwin, suggested an animal origin, these three illustrations do more than just document an observation.[87] Instead, these types of images—from 1845, 1859, and 1871—demonstrate the interaction of the forces that combine to produce evolution: chance variation and continuous selection. Only illustrations can visually present the interplay of these two factors over thousands of generations—a period of time vastly beyond the horizon of human experience. A process that in real time surpasses the bounds of any human life, and thus the capabilities of any human individual, can be taken in at a glance: the accumulation of minor changes over millennia appears in the slant of a line or as a sequence of alterations in a picture series.

A series of interventions and translations was required before the argus pheasant's ornament could appear in a picture series in *Descent of Man:* an argus pheasant in the Malaysian tropics became a hunting trophy; the trophy became a specimen in a museum; the specimen was captured in a series of sketches; and the sketches became printing proofs. These in turn formed the basis for the woodcuts included within the book's pages that together constitute the series itself.[88] Each step in the ornament's journey from bird's feather in Malaysia to the page of an English book required the work of specialists, from colonial traders and sailors to hunters, taxidermists, collectors, and museum employees, and finally to professional illustrators and printers. His way of looking at the argus pheasant's plumage had developed historically as well: since the 1840s Darwin had examined earlier picture series from the fields of anatomy, physiognomy, and embryology, merging their approaches and finally applying this new perspective to the argus pheasant.

What is surprising about this process is the simplicity of the means Darwin used to pull together this network of specimens, notes, pictures, illustrations, and sketches. His workspace consisted of a few sheets of light-blue paper, and ink and lead pencil were the only materials he used to record his thoughts. In a few strokes, his drawing hand compiled natural history's pictorial traditions into a collage, rearranging them and adding new elements. The lack of artistic ability to which Darwin himself admits—his "incapacity to draw"—in no way detracts from the sketches. In contrast to Agassiz, Huxley, or Haeckel, who were all excellent artists, Darwin did not possess the talent to eliminate nature's imperfections on paper.[89] It was precisely this inability to overcome nature's inadequacies that proved how well Darwin's eye and hand worked together. His gaze remained stubbornly fixed on nature's flaws; like a seismograph, his hand traced the breakdown of natural theology and teleology in shaky lines, wavy outlines, and irregular compositions to create a new image of nature. After Darwin, nothing in nature could have been created without chance: no organism and no organ was perfect. All living things existed in constant flux, never arriving at a state of perfection or completeness. The diagram in *Origin of Species* showed that evolutionary history followed a zigzag course, while the picture series in *Descent of Man* came together in fits and starts to culminate in an image of imperfection. Nature's products were not works of art, nature was no artist, and the eye that gazed upon it was not a mechanical instrument— Darwin's imperfect style of drawing perfectly corresponded with the observations it recorded.

As in the case of the diagram, few of Darwin's successors broke with the tradition of natural theology in the way that Darwin's pictures had. Haeckel's *Anthropogenie*, which, like the second edition of Darwin's *Descent of Man* appeared in 1874, also begins with a frontispiece featuring a picture series (Figure 53). A being in an elementary embryological state develops over a series of four images into a mammal to demonstrate—thanks to their similarity in this early stage—the family ties between humans, sheep, bats, and cats. In quick succession the pictures move from simple original form to a state of completion, and the evolution of the human face culminates—as it did in Johann

Figure 53. Frontispiece to Haeckel's 1874 *Anthropogenie, oder Entwickelungsgeschichte des Menschen* (Anthropogenie, Or the Evolution of Man)

Caspar Lavater's drawing—with the image of a god (this time Zeus rather than Apollo). Just as Haeckel reincorporated the old ladder-based diagram into evolutionary theory when he portrayed the evolutionary family tree in *Anthropogenie* as an adult oak, he returned in the frontispiece to the very visual tradition that Darwin had left behind. Darwin had wrung a way to make chance visible by incorporating disorder, breaks, and the accidental into his drawings. In Haeckel's work, all these hard-won traces of chance were swept away. In the wake of Darwin's picture series, which was scattered across several pages in *Descent of Man*, Haeckel's illustrations regroup on a single plate to follow the direction of time and end at a state of perfection. Haeckel's ornamental style did not allow for evolution's preliminary, imperfect, and accidental aspects. Once again, nature came to resemble the work of art that Darwin, with his aesthetic of chance, had sought to overturn. But according to Francis, one of Darwin's five sons, the feather with the imperfect eye-shaped ornament became one of his father's favorite objects—one that he used to explain evolution to his visitors in Downe.[90]

4 THE LAUGHING MONKEY

The Human Animal

O N April 21, 1868, Darwin sketched a family tree in brown ink on a white sheet of paper—a spindly image measuring less than five inches (Figure 54 [Color Plate 4b]). At the bottom of the sheet he noted that it was part of the "primates" family tree—in other words, the division of the animal kingdom to which humans also belong. The image evokes the feeling of being witness to a mighty mental struggle. Deletions and corrections overlap one another; the sketch is composed of multiple layers, the fragments of discarded ideas. In the upper forks of the diagram the first draft shows through in tentative broken lines, which were later copied over with solid strokes. Darwin hesitantly indicated a number of branches as if they were questions and then painted over them with thick lines. He drew other branches that trail off into oblivion and remained unnamed. Where a name is present, it is often preceded by an earlier one that has been crossed out.

Despite this struggle, Darwin apparently arrived at a conclusion. Read from the bottom to the top, the family tree tells the following story: the earliest branch of the primates in the history of evolution is the *Lemuridae*, a kind of half-ape whose descendants live today on the African island of Madagascar. Then the family of the apes splits into the lines of the "new world monkey" and "old world monkey." An unnamed lineage at the very left

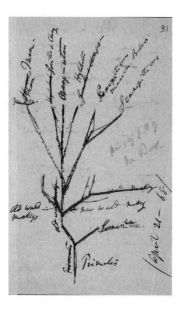

Figure 54. Darwin's 1868 family tree of the primates (Archive [DAR 84.91], with the permission of the Cambridge University Library Syndicate) (Color Plate 4b)

trails off and dies out. This dead end is followed by three major branches: humans on the left; the great apes with the genera of gorillas, chimpanzees, orangutans, and gibbons in the middle; and the lesser apes on the right, represented by the two genera of *semnopithecus*. The evolutionary diagram thus makes two statements about human beings: first, that gorillas and chimpanzees are our closest living relatives; and second, that humans represent just one part of the primate family. Darwin grants humans a position that is neither central nor exceptional. As a result, his scheme clearly differs from that of the German zoologist Ernst Haeckel, whose 1874 *Anthropogenie oder Entwickelungsgeschichte des Menschen* (Anthropogenie, Or the Evolution of Man) would give humans a central place at the very top of the family tree, banishing the gorilla and chimpanzee to the lower branches (see Figure 35). Darwin's plan also differs from Haeckel's in that it remained unpublished. He neglected to include the 1868 drawing three years later in *The Descent of Man*, four years later in *The Expression of the Emotions in Man and Animal*, or in the revised editions of *On the Origin of Species* that appeared during the 1870s. He continued to observe the lesson he had learned from his reading of Chambers: to avoid speculating about genealogies in his own work. The abstract chart of evolution, bound into *On the Origin of Species* as a fold-out table in 1859, remained Darwin's only published evolutionary diagram, and he kept the fact that he considered gorillas and chimpanzees to be our closest relatives to himself. Darwin's reticence in this case is another example of the pattern of silence examined throughout this book. Many have taken note of his years of refusal to comment on the question of whether humans are part of the history of evolution. He made no mention of evolution in the account of his travels, he omitted humans from the founding work of evolutionary theory in 1859, and later he avoided the question of what humankind's more immediate family tree must be like. However, none of these silences is so telling as the one he maintained about the gorilla in *The Expression of Emotions*. In the 1860s the gorilla—our closest relative in Darwin's sketch—had come to embody the essence of bestiality.

The stylization of the great apes into monsters in the 1860s was a result of the publication of *Explorations and Adventures in Equatorial Africa*, an account of his travels by the explorer Paul Belloni du Chaillu. Du Chaillu returned from his African journey, which was financed by the Academies of Natural Sciences in Philadelphia and Boston, the same year that *On the Origin of Species* appeared in England. He claimed to have seen gorillas in the wild a number of times and was thus considered the first significant eyewitness observer of this otherwise legendary animal. Zoologists had only become aware of the gorilla's existence in 1847, when the species was described based on two skeletons in the Natural History Museum in Boston and distinguished from the orangutans and chimpanzees,[1] which were already known. Before Du Chaillu, no Western traveler had ever seen the animal.

In England, *Explorations and Adventures in Equatorial Africa* was published by Murray in London—the same house that published Darwin's books—in 1861. The frontispiece to Du Chaillu's report clearly shows what was at stake should humanity be forced to accept the gorilla as a relative (Figure 55). A male gorilla standing to its full height—dark, shaggy, and frighteningly large—strides toward the viewer. The beast's right leg is braced on a tree root, while next to the upright left leg hangs an arm that is twice as long as the leg itself. With its right arm, the animal grasps a branch and holds itself upright. The viewer can make out a muscle-bound rib cage under the fur that terminates at the gorilla's almost cubical skull. The animal has almost no neck; its head is turned to the side and the face sits atop a thick bulge of flesh. Wrinkles begin at the small eyes and follow the sides of the flat nose to the maw below, which, despite its considerable size, cannot quite contain the huge incisors. The lower lip hangs forward as if it were made of elongated rubber, seemingly unable to close over the fangs. The picture implies that such an animal tirelessly emits threatening sounds, and in fact the gorillas Du Chaillu reports meeting are almost always bellowing. At his first encounter it is the beast's "devilish cry" that draws his attention to the gorilla in the first place, while at the second meeting the animal approaches the hunter with a "hideous roar" and beats its chest with its fists.[2] When it comes to anthropoid apes, Du Chaillu has only terrible things to report. To him, even a three-year-old male gorilla that survived for just a few days after its mother was shot by hunters seems devious and evil. For Du Chaillu, the animal's unsavory character is revealed by the "wicked" look in its eyes and its "morose" and "ill-tempered" expression.[3] Like Darwin, who would write *The Expression of the Emotions in Man and Animal*, Du Chaillu believed that animals' faces could communicate, but the American's interpretations of their expressions transform a feeling of familiarity into a divide. Although the gorilla's similarity to a human visage enables humans to read and understand the animal's face, what Du

Figure 55. Frontispiece for Du Chaillu's 1861 *Explorations and Adventures in Equatorial Africa*

Chaillu sees there leads him to conclude that the gorilla is inhuman after all.

The gorilla's teeth-gnashing appearance also spelled the end of any nostalgic dreams of Eden the viewer may have harbored (Figure 56). This image has striking similarities to Albrecht Dürer's *Adam and Eve* from 1504, including the poses of Adam and the gorilla, the way their genitals are hidden by a branch, and the depiction of the forest and the clearing. Resemblances with depictions such as this one raise the question of whether the ape should be viewed as the "new Adam": just over a year after Huxley and Wilberforce had publicly debated whether humanity's

Figure 56. Albrecht Dürer's *Adam and Eve*, 1504

ancestors resembled the apes, this most terrible of figures appeared and seemed to drive mankind from Paradise, replacing
the nude human form with a black, hairy horror. The thesis that
humans were related to this animal was taken to mean that
mankind had been born from a monster possessing all the traits

that white Europeans loathed: it was cruel, stupid, hairy, and black.[4] Darwin responded to this image in *Expression of Emotions,* his last major work. Surprisingly but nonetheless clearly, he addressed the issue of our kinship with animals by approaching it from the opposite perspective. His critics had argued that the theory of evolution would dangerously undermine the concept of "human," but Darwin left the image of humanity untouched. Instead, he showed how his theory fundamentally redefined our perception of animals, leaving no opportunity to dismiss them as "monsters." This would be the last time he would write on this subject. Once the book was finished, he told Haeckel: "I have resumed some old botanical work, and perhaps I shall never again attempt to discuss theoretical views." From that point to his death in 1882 he devoted his writings to plants, insects, and earthworms.[5]

THE PHYSIOGNOMIC TRADITION

Expressions of Emotions contains more illustrations than any of Darwin's other books. His only work to feature photographs or include pictures of people, it initially was also Darwin's most commercially successful work, selling nine thousand copies in England in the first four months. By the turn of the century, it had also appeared in the United States, Holland, France, Germany, Italy, and Russia.[6] Examining the basis of human and animal expressions was a necessary step in Darwin's work on evolutionary theory, and one he took immediately after completing *Descent of Man* in 1871. In that book he had already described how the similarity he postulated between human and animal was not just a matter of physiology or anatomy, but that it included aspects of our spirit, essence, and understanding. In the third chapter, he summarizes:

> It has, I think, now been shewn that man and the higher animals, especially the Primates, have some few instincts in common. All have the same senses, intuitions, and sensations,—similar

passions, affections, and emotions, even the more complex ones, such as jealousy, suspicion, emulation, gratitude, and magnanimity; they practise deceit and are revengeful; they are sometimes susceptible to ridicule, and even have a sense of humour; they feel wonder and curiosity; they possess the same faculties of imitation, attention, deliberation, choice, memory, imagination, the association of ideas, and reason, though in very different degrees. The individuals of the same species graduate in intellect from absolute imbecility to high excellence. They are also liable to insanity, though far less often than in the case of man.[7]

The key point here is the choice of traits that, according to Darwin, human and animals share—emotions such as jealousy, suspicion, ambition, gratitude, generosity; capacities such as humor, wonder, and curiosity; abilities such as imitation, attention, consideration, choice, memory, imagination, and the association of ideas; and even madness. Interestingly, items with positive connotations clearly dominate. Darwin considers humanity's inheritance from the animals to be a largely positive one that even included—as the previous chapter shows—an aesthetic sense. The idea that this essential kinship similarly manifested itself in both humans and animals formed the basis of *The Expression of the Emotions,* the second work to follow *On the Origin of Species.*

Darwin wrote *Expression of Emotions* within four months of the date he submitted the proofs for *Descent of Man*—an assembly-line pace that reveals how closely the two works are related. He had originally planned to publish this research as part of *Descent of Man* but soon realized that he had enough material for a separate book. *Descent of Man* illustrated the inner lives of humans and animals in numerous examples until the two overlapped and could no longer be distinguished. In *Expression of Emotions* he examined these depictions in light of the system of evolution that underlies the ways in which this inner life took on outer expression. Human and animal faces merged to an indivisible whole: in the pages of *Expression of Emotions* the reader encounters apes who laugh and ladies who snarl.[8]

Seen through the lens of evolutionary theory, neither behavior is surprising.

Darwin believed that facial expressions and gestures among animals and humans were caused by both habit and nervous impulses. According to this view, expressions that do not fulfill a function could still be commonly used if they had become habitual—snarling in anger, for example, was a habitual threatening gesture derived from the action of biting. According to Darwin's theory, other expressions, such as laughing or crying, were directly linked to nervous stimulation in that they provided relief from states of nervous excitement. But whether it was the result of habit or whether it served a functional purpose, each expression and gesture had a history. It had developed in the course of evolution just as the Galápagos finches, the argus pheasant's ornament, or the human eye had done. In *Expression of Emotions* Darwin tried to reconstruct this history. He considered children, animals, and the insane to be particularly amenable to study because, as he claimed, they expressed their feelings in an uncontrolled and direct way.[9] To gain an understanding of the purpose and function of particular gestures, Darwin corresponded with a number of physiologists, such as Frans Cornelis Donders in Utrecht, or consulted their works, such as those of England's Charles Bell or France's Guillaume-Benjamin Duchenne de Boulogne. For example, Darwin commissioned Donders to perform experiments to discover why humans emit tears. Donders concluded that crying was a protective mechanism of the eye.[10] But before providing physiological explanations for various expressions, it was necessary to demonstrate how they manifested themselves in the first place. Do apes cry? How does a rabbit show fear? Which animals experience their hair standing on end? Do swans' feathers stand up too? Can dogs smile? To study the meaning of these manifestations Darwin needed to identify the expressions that animals actually demonstrate, and for this effort he accumulated an extensive collection of pictures.

The many notes on the expressions of humans and animals in his early diaries show that his interest in the subject was as old as his work on evolutionary theory itself. In 1838 Darwin had already compiled an extensive list of questions on the subject in his notebook, asking, "What is Emotion analysis of expression of desire—is there not a protrusion of chin, like bulls & horses. . . . How is expression of anger in species of swans, in parrots &c &c—peacock & turkey cock in passion."[11] At one point he wrote, "Seeing a dog & horse & man yawn makes me feel how all animals built on one structure." In the Zoological Garden he observed the orangutan pouting, the "pig-tailed baboon" being "sulky," and the chimpanzee Jenny hiding under a blanket out of "fear or shame." On September 16, 1838, he noted: "Endeavored to classify the expressions of monkeys." When his first child, William, was born in 1839, he followed his son's development, noting his first expressions and gestures in his notebook as well.[12]

The notebooks also contain references to potential predecessors in the physiognomic tradition, especially to the *Fragments of Physiognomy* of Johann Caspar Lavater. At the age of twenty-nine, Darwin wrote in Notebook N, "Seeing some drawings in Lavater, P. cii Vol III of excessively cross-half furious faces which may be described as an exaggerated habitual sneer the manner in which whole skin or muscles are contracted between eyes & upper lip, is most clearly analogous to a panther I saw in garden uncovering its teeth to bite."[13] As we saw in the previous chapter, Darwin owned the ten-volume French edition of Lavater's *Fragments,* which contained the twenty-four images of a frog metamorphosing into Apollo. In England, Lavater's writings were widely known and among the standard works sold by book clubs during the nineteenth century. In 1810, more than thirty years after the book first appeared in German, twenty different translations were available. Darwin even experienced firsthand how strongly Lavater's idea that facial features reveal character traits had taken hold. In his autobiography he recounts how his nose nearly cost him his place on the

H.M.S. *Beagle*—Captain FitzRoy, a follower of Lavater's, believed its shape indicated a lack of "energy and determination." When Darwin returned from his voyage around the world, his father reportedly cried out: "Why, the shape of his head is quite altered."[14] His son's account suggests that the elder Darwin greeted this development as a change for the better.

While the pictures in Lavater's books, such as the aforementioned face drawn in rage, held a certain fascination for Darwin, he cared less for the labyrinthine texts. "Is there—anything in these absurd ideas?" he wrote in Notebook N. "Whole volume skimmed," he noted at one point in his edition of Lavater's work, and at another, "If I want to show what rubbish has been written a translation of this will do."[15] His annoyance and his skimming of Lavater's study point to the differences between his ideas and Lavater's theory of physiognomy. The most obvious point of distinction between the two projects is that Lavater, who may have shown humans and animals to be similar, never intended to claim that they were related. A human resembled a cat not because both had common ancestors, but rather because he shared the cat's unpredictable wildness.

Similarity to a cat had nothing to do with heredity but rather with character traits, which Lavater believed could be found as pure types in animals. He did not examine physiognomies hoping to find history and kinship; instead, he deciphered what he believed was the signifying power of a universal language of nature that shaped the expressions of all creatures.[16] In Lavater's mind, animals stood for defined attributes, and just as slyness in traditional fables takes the form of a fox, the face of a sly person resembles that of a fox as well. Lavater hoped that the systematic study of human and animal physiognomy would make the human visage legible, so that the nose, eyes, mouth, and cheeks would reveal the character of their owner. The faces of "lions, tigers, cats, leopards," for example, show "shiftiness, disloyalty, and ferocity." The elephant, in turn, epitomizes memory and cleverness, and thus a head with similarly proportioned forehead promised "deep, superior judgment." In its lack of ambiguity, the animal face could serve as a key for deciphering the traits hidden in the human countenance.[17]

Darwin, however, did not see the faces of animals as simplified guides to human ones. Since humans and animals were related, it was not possible to determine whether the human face could be used to decipher that of an animal, or vice versa. As a result, Darwin acknowledged an expression of "a thoroughly canine snarl" in the photograph of a woman pulling back her upper lip to show her incisors, but he likewise designated the movements of the muscles around a crested macaque's mouth and eyes as "smiling" or "laughing"—emotional responses that, until then, had been considered exclusively human.[18] The face of the dog thus helped him to understand the woman's expression, but the human face also allowed him to read that of the ape. In his view, an animal's expressions are just as human as a human's are animalistic. Lavater attempted to systematize the countless appearances human and animal faces could take and derive an underlying rule from this wealth of examples in a typological approach that, as scientific historian Michael Neve has shown, was in direct opposition to Darwin's system of evolution. For Darwin, nature's fluid boundaries and constantly branching variations played the same role in determining expressions, faces, and gestures as they did for other physical attributes. Set, unchanging characteristics were as inconceivable in this regard as in any other.[19]As a result, neither the woman nor the ape serves as a representative of the species to which each belongs. The ape that artist Joseph Wolf observed for Darwin did not prove that all apes laugh, but only that this particular individual had this ability; the same is true for the woman and her snarl. The pictures show individuals, not types. It is therefore not surprising that the picture archive Darwin assembled during his research—which will be discussed below—does not contain any illustrations by Lavater.

A work that was more significant in Darwin's thinking was *The Anatomy and Philosophy of Expression as Connected with the Fine Arts,* by the Scottish physician Sir Charles Bell, a book that is largely forgotten today. Bell, like Darwin a graduate of Cambridge, first published his work in 1806. He wrote primarily for artists, but this intended readership seems unlikely to have responded positively to the author's brusque tone.[20] Bell hoped to keep artists of the new generation from repeating the mistakes of their predecessors. He believed that artists clung too closely to tradition, studying the works of antiquity instead of those of nature. The depiction of animals in the works of the old masters was a particular thorn in his side; he criticized Giulio Romano, a sixteenth-century Italian mannerist, for the horses his heroes rode into battle, and Peter Paul Rubens for the lions surrounding his Daniel in the lion's den. One target of his displeasure was the furrowed brow of the king of beasts, which lent the animal a kingly air but which was anatomically impossible. The good doctor's objection to Romano's battle steeds was also based on their supposed inaccuracy. The artist had shown the horses with open mouths to depict their agitation, but Bell protested that "in the utmost excitement, animals of this class do not open the mouth; they cannot breathe through the mouth,—a valve in the throat prevents it—."[21]

Point by point, Bell listed the mistakes he had discovered in works of art to dispel the human mask he felt had been imposed on animals. For Bell there was a simple reason why animal expressions departed so much from those of humans: the muscles under their faces were utterly different. He included a number of cross-section illustrations to prove this contention. Whether lion, horse, or dog, the result was the same—Bell identified eighteen different muscle groups in the human face, but never more than fifteen in that of an animal.[22] The key difference was the lack of the forehead muscle, *Musculus corrugator,* which controls brow and eye movements in humans. For Bell there was a simple explanation for this difference: along with language, the human face had been created by God so that humans could communicate with one another. Animals had no more need for communicative gestures than they did for language itself. As a result, Bell argued, the expressions animals could make were limited; their faces were "chiefly capable of expressing rage and fear." His teleological explanation of the facial muscles as something created by God to fulfill a purpose reveals him to be an adherent of the English school of natural theology, and he provided commentary for later editions of Reverend William Paley's book on this subject (see Chapter 3).[23]

But even Darwin's observations in his early notebooks contradicted Bell's theory. Before he learned that animals supposedly lacked a *Musculus corrugator,* he had seen the orangutan in the zoo pouting and had claimed the dog at home laughed "for joy." Between 1838 and 1839 he filled two notebooks with his observation of animals; in 1840 he read Bell—and came to the conclusion that the doctor was mistaken.[24] With his own eyes he had seen the faces of animals take on expressions that, in Bell's analysis, were impossible. Bell wrote that animals lack the muscle to wrinkle their brows, but Darwin had noted: "I have seen well developed in monkeys incessantly clenching skin over eyes." Bell also viewed smiling as a purely human movement, one "bestowed to express the affection of the heart."[25] Darwin could not accept that human facial muscles had been created for the sole purpose of communicating our feelings to other people, as he wrote to Wallace in March 1867: "I want, anyhow, to upset Sir C. Bell's view . . . that certain muscles have been given to man solely that he may reveal to other men his feelings."[26]

The method Darwin employed to research his subject was diametrically opposed to Bell's approach. As a physiologist, Bell had dissected human and animal faces and drawn conclusions about their expressive ability based on the anatomy of their muscles. Darwin did the opposite: by carefully observing and recording facial movements, he determined which muscle groups were involved. From what lay under the skin, Bell determined what could happen on the surface, while Darwin studied the surface level to determine what lay below. The discussion of the subject from an anatomical perspective in *Expression of Emotions* is correspondingly brief. While the beginning of the book included three reduced engravings showing the human facial muscles according to Bell and the German anatomist Jacob Henle, Darwin cautioned that they provided a general orientation only, since "the facial muscles blend much together, and . . . hardly appear on a dissected face so distinct as they are here represented."[27] For this reason, he continues, even anatomists do not agree on how the muscles are linked; opinions on the number of muscle groups range from nineteen to fifty-five, but in their flexibility and arrangement "they are hardly alike in half-a-dozen subjects."[28] Some people can raise their upper lips on one side, while others cannot; there are people capable of moving their nostrils and others who are not. Capturing the expression of emotions, therefore, requires looking at the surface, not under it. Darwin's approach to the problem resembled that of physiologist Guillaume-Benjamin Duchenne de Boulogne, one of his few scientific predecessors whose studies he valued.

PEOPLE IN PHOTOGRAPHS

Beginning in the mid-1860s Darwin systematically compiled a collection of two hundred engravings and photographs that formed an archive of human and animal behaviors and expressions. This archive is located today in two portfolios in the Cambridge University Library. The material they contain is astoundingly diverse: along with a photographic reproduction by Innsbruck's F. R. Bopp studio of an oil painting showing the "hairy family of Ambras" we find a photo of a melancholy women from the Wakefield insane asylum, a studio portrait of a crying baby, and an engraving of a panther lurking on a cliff for its prey. A print of José de Riberas's praying *Mary of Egypt* from the Picture Gallery in Dresden lies together with whinnying horses, and a number of other horses, a cat arching its back, and a gentleman donning a hat all make an appearance. Darwin's sources for these images are no less varied. The works come from studio photographers such as Oscar Gustav Rejlander, Adolph Diedrich Kindermann, and Giacomo Brogi; the animal

artists Sir Edwin Landseer, Joseph Wolf, and Briton Rivière; the
London Stereoscopic Society; the Collection Anthropologique
du Muséum de Paris; James Crichton-Browne, the director of an
insane asylum; and the famous French physiologist Guillaume-
Benjamin Duchenne de Boulogne. Only a fraction of this collection found its way into the published version of *Expression of
Emotions.* The art historian Phillip Prodger, who viewed and
catalogued the material in Cambridge, furthermore suspects
that even more material originally existed than what has survived in the archives.

Darwin spared no effort in collecting material to ensure that
his inventory of the forms of expression was as complete as possible. Now that he had finally decided to write his long-awaited
book on human evolution, he began leaving his country home in
Downe more frequently. His two horses—a white and a chestnut,
whose expressions also served as his objects of study—were har-
nessed to his coach for the six-mile drive to Bromley Station,
where he caught the train for a forty-minute ride to Victoria Station in London. As Prodger has shown, Darwin embarked on extended forays through the capital's book and photography shops.
"Like a geologist armed with a rock hammer," Prodger writes,
"Darwin mined the shops of London searching for useful fragments of expressive imagery."[29] He traveled through stores from
Victoria to Baker Street and from the Strand to South Kensington, and also spent many hours in the Zoological Garden in Regent's Park; he conversed with attendants, sent questionnaires to
correspondents around the world, questioned mothers, and studied his own children and pets, especially the little dog Polly whose
basket was kept in his study.[30] Some of the finds from his trips to
London appeared later as illustrations in *Expression of the Emotions,* and he commissioned others from the animal painters he
met at the zoo to create additional pictures for his book.

Darwin had discovered photography—a young medium not
widely used in science—during a vacation on the Isle of Wight
in the summer of 1868.[31] There he met the well-known Victorian
photographer Julia Margaret Cameron when the Darwin family

rented the photographer's vacation home. Influenced by the Pre-Raphaelites, Cameron began taking pictures in the 1860s, concentrating on scenes of tender intimacy that straddled the border between portraits and genre studies. Her work clearly shows her interest in staging, and she often had her models re-create historical or mythical events in so-called *tableaux vivants*. In one instance Horace Darwin, one of Darwin's sons, was put to use posing for a scene of King Arthur's resurrection along with the servants, who were dressed as knights, princes, ladies, and lackeys. Other photographs had only vague connections to mythology, such as that of a little girl equipped with wings who resembles the putti in Raphael's *Sistine Madonna*. Cameron often used young girls as models, a preference she shared with many Victorian studio photographers. The portraits of children by the writer Lewis Carroll, for example, were well known, and Darwin himself owned an example—a print of a smiling girl with the title *No Lesson Today*.[32]

During the family vacation on the island, Cameron photographed Darwin as well. In 1867 she had captured the likeness of John Herschel, the astronomer who had dismissed the theory of evolution as "the law of the higgledy-piggledy." Depicting him with fluttering white hair, she created a figure in the mold of a Byronic hero. Cameron lit Darwin from above for the 1868 portrait, accentuating his forehead, beard, and brow-shadowed eyes (Figure 57). Both his family and acquaintances were struck by the resulting theatrical effect. Darwin's sons were reminded of Moses, as was the botanist Joseph Hooker, a friend who had seen an earlier photograph of Darwin with a beard and written, "Glorified friend! Your photograph tells me where Herbert got his Moses for the fresco in the House of Lords—horns & halo & all."[33] For his part, Darwin was pleased with the portrait. He paid Cameron four pounds seven shillings and later ordered additional copies, which he signed and sent to his correspondents.[34] This portrait also served as the model for the 1871 caricature from *Fun* (see Introduction). Most recently it was copied in 2000, when Darwin's picture replaced that of Dickens on the English ten-pound note.

Figure 57. Julia Margaret Cameron's 1868 portrait of Darwin used on a visiting card

Figure 58. Oscar Rejlander's 1857 tableau vivant, *The Two Ways of Life*

Another photographer whom Darwin knew personally, and who eventually proved to be the most important contributor to *Expression of Emotions*, was Oscar Gustav Rejlander. Rejlander had introduced Cameron to photography in 1863; Darwin, in turn, met him in London in 1871. Obvious parallels existed between the two photographers: like Cameron, Rejlander produced popular portraits of children and was fond of lavish tableaux vivants. Educated as an artist in Rome, he first used photography as an aid for painting before discovering it as an artistic medium in its own right. He worked on the monumental *Two Ways of Life* (Figure 58) for six weeks, piecing together more than thirty negatives to create the final image. The result, a lavish photographic allegory, was presented in 1857 at the Art Treasures Exhibition in Manchester.[35] Emulating populous works of art such as Raphael's *School of Athens*, Rejlander squeezed almost thirty models into his composition. On the left-hand side, personified vice lolls about in the form of barely clothed female visitors at an Oriental bath, while on the right the virtuous pray, weave, and hammer. Two youths stand in the middle, faced with the choice of which path to take. Because six exposed female breasts were visible in the composite photograph, prints had to be partially shielded from view at an exhibition in Scotland. This raciness, however, did not keep Queen Victoria from purchasing the work for her husband, Albert.

Rejlander provided twenty of the images published in *Expression of Emotions*, more than half of the photographs published in the book. He was also the only contributor who appears himself as a model. Rejlander often appeared within his own work; in *The Two Ways of Life*, for example, he donned an antique robe to guide one youth to the straight and narrow path while casting the other a warning gesture to avoid the path of vice. In *Expression of Emotions* he can be seen with his hat, beard, and velvet jacket demonstrating the concepts "annoyed," "helpless," "shrugging the shoulders," "disgusted," "defending himself," and "frightened." The snarling woman who exposes her incisor in an expression of contempt is Rejlander's wife. Just as Rejlander had earlier considered photography as an extension of his painting career, he now seemed to view working for Darwin as a way to further his activities as a studio photographer. None of the other contributors developed such a trusting relationship with their client. On occasion Rejlander even sent photographs to Darwin for no other purpose than to share a joke.[36] The father of evolutionary theory wanted nothing from him beyond what he was doing already; the photographer was even free to continue using costumes and props. As a result, Rejlander wears a hat in some shots, while in others his hair stands disheveled on end. Usually he sports his velvet housecoat as well.

These seemingly marginal details that create a homey atmosphere are key to understanding *Expression of Emotions*. Darwin's correspondence with his publisher reveals the importance he placed on the photographs. The images were included at his express wish and only after tireless argument with John Murray, who feared that seventy-five pounds for a thousand prints would cut too deeply into any profits. In addition to increasing costs, halftone processes were largely unknown territory—a further obstacle Darwin was willing to accept.[37] In its first edition, *Expression of Emotions* contained seven plates, each made up of two to seven images. Each plate corresponds to a chapter that examines the spectrum of a particular category of emotions. One chapter focuses on "Hatred and Anger," for example, while the subject of another is "Surprise, Astonishment, Fear, Horror." Contrary to what readers might expect, animals and humans are presented separately in the illustrations to *Expression of Emotions* rather than in side-by-side comparisons.

More than half of the models in the book are children. This preponderance is partly because Darwin considered children to be particularly suited for such studies of expression. It is also no coincidence that such child portraits were one of Victorian photography's most beloved genres. The taste of the age, which can be seen in almost all the photographs included, is most clear in the preference for portraits—the most common application of the medium in its early years. One sign of this tradition is the oval passe-partout framing of the photographs on many of the plates. In the history of portrait painting, this oval format was most common for less representative paintings intended for private use.[38] The intimate quality of such pictures encourages the viewer to imagine the stories behind the reactions they document.

"Extreme disgust," Darwin writes, commenting on Plate V, "is expressed by movements round the mouth identical with those preparatory to the act of vomiting." Originally, he explains, disgust developed in relation to eating, and the expression it provokes is therefore most pronounced around the mouth even when the emotion is prompted by factors other than food. Expressions of disgust thus represent a rudimentary form of nausea—a physiological reaction through which the body prevents the consumption of spoiled food. "The mouth is opened widely," Darwin explains, "with the upper lip strongly retracted, which wrinkles the sides of the nose, and with the lower lip protruded and everted as much as possible."[39] One example of disgust on Plate V is an oval-framed portrait of a young woman whose light-colored blouse is closed with a brooch high on the neck (Figure 59). "It represents a young lady," Darwin comments, "who is supposed to be tearing up the photograph of a despised lover."

Figure 59. Plate V, contained in the chapter about "disdain, contempt, disgust, guilt, etc." in the 1872 *Expression of Emotions;* Oscar Gustav Rejlander appears at the lower left and right

The reader would have been reminded of the illustrations to Victorian novels like those Charles Dickens described in his weekly magazine *All the Year Round.* One of the most popular motifs of such pictures, Dickens wrote, was "young ladies reading love-letters, or overwhelmed with some piece of ill news just received."[40]

Surprisingly, historians so far have been most interested in those photographs that are least typical for *Expression of*

Emotions—the ones Darwin took from the 1862 study *Mécanisme de la physionomie humaine ou analyse électro-physiologique,* by the French physiologist Guillaume-Benjamin Duchenne de Boulogne. Darwin owned the book and included eight pictures from it in *Expression of Emotions:* two of actors hired by Duchenne and six others of an old man the physiologist referred to as the "irritable cadavre."[41] To the latter group Darwin made certain changes that underscore the character of his book and distinguish his views from Duchenne's. Like Darwin, the French scientist believed that understanding facial expressions required studying muscular movements on the surface of the face. For this reason, Duchenne included eighty-four plates in *Mécanisme,* some of which documented several expressions. To study these fleeting phenomena, Duchenne developed a laboratory experiment in which the skin of his subject's face was stimulated with electrical current (Figure 60). The guinea pig in this case was an old man—the aforementioned irritable cadavre—who owed his macabre moniker to a disability that had left him with practically no feeling in his face. From the researcher's perspective, this unique quality made the elderly man an outstanding object of study, since no sensations of pain would cloud the results of the experiment. Duchenne attached two electrodes to the skin of the man's face and stimulated the underlying muscles, which contracted in response to the electric current. The movements that resulted were as involuntary as the twitching of the frog's legs in Alessandro Volta's famous electrical experiments in Como in the eighteenth century. The old man's expressions were also completely superficial, since they were prompted purely by the electrodes, not by any emotional state. Without actually experiencing joy, happiness, sorrow, anger, or fear, the subject responded to the current with folds of laughter, starting eyes, or bulging neck muscles. Thanks to the steady flow of electricity, the expressions could be fixed long enough so that even the cameras of the day, with their long exposure times, could capture them. Duchenne and his employee Adrien Tournachon, the brother of the famous French photographer Félix Nadar,

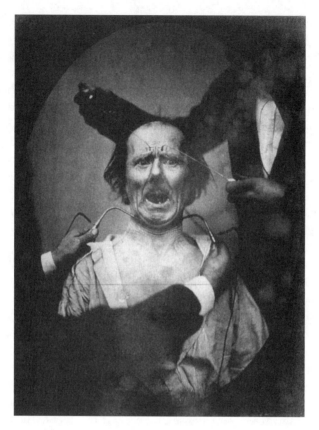

Figure 60. Photograph from Duchenne's 1862 *Mécanisme de la Physionomie Humaine*

recorded the old man on film as his face involuntarily grimaced and twitched.[42] Little is know about this notable research subject beyond his monstrous nickname and the photographs themselves.

Initially, Darwin saw the same value in Duchenne's work that the physiologist did himself: the use of photography as a recording mechanism, the artificial induction of facial expressions, the experimental stimulation of parts of the face. Duchenne had shown that the facial muscles always move in groups, so that the muscular anatomy at rest can provide little indication of the expressions it generates. Instead, the face can be studied only in motion and from the surface. Again like Duchenne, Darwin was interested in the ambiguity of human expressions and in countering Bell's claims that the ultimate purpose of expressions was communication and that only humans possessed this ability. If the messages our faces communicate are ambiguous, it is unlikely that God had created expressions for the purpose of communication, as Bell had suggested. Both Duchenne and Darwin showed the photographs to random individuals, and both reported that the same expressions were interpreted in various ways.[43] Despite these similarities, the differences between the two men's approaches are key to understanding Darwin's studies of expression. Underlined passages in his copy of Duchenne's book suggest that he thoroughly evaluated Duchenne's photographs in terms of the facial movement they portrayed. However, he still made changes to them for the readers of *Expression of Emotions*. For example, oval passe-partouts almost completely obscure Duchenne's electrical machinery on Plate III, where, in the chapter on "Joy, High Spirits, Love, Tender Feelings, Devotion," the old man makes his first appearance (Figure 61). On the left-hand side are the faces of three girls, while on the right the old man appears, from top to bottom, "in his usual passive condition," then "naturally smiling," and finally under the influence of electricity "with the corners of his mouth strongly retracted by the galvanization of the great zygomatic muscles." Darwin's commentary refers to this final expression as the "false smile." In Duchenne's experiment, only the lower part of the mouth was stimulated, without the involvement of the eyes and upper lip visible in the natural smile. Thanks to the incomplete nature of the induced expression, Darwin reported, three of the twenty-four people who were shown the photo could not say which emotion it represented at all, while the others replied in vague terms such as "'a wicked joke,' 'trying to laugh,' 'grinning laughter,' 'half-amazed laughter, &c.'"[44]

Figure 61. Plate III from *Expression of Emotions* showing the act of smiling

Darwin, as we have seen, flanked the old man with pictures of three girls. These shots all come from the London photo studios of George Charles Wallich and Rejlander and show, according to Darwin, "different degrees of moderate laughter and smiling": propped on an arm, wearing a hat or a flounced dress, the girls smile with sparkling eyes into the camera.[45] The "moderate laughter" of the girls illustrates why the old man is indispensable. As pleasant figures that themselves evoke smiles, the girls are eminently suitable for demonstrating this expression. But their fresh childish faces fail to cover the expressive spectrum. At least two parts of the face—the muscle groups around the mouth and those that control the eyes and forehead—are involved in a smile. But on the girl's faces, only the movement of the mouth can be seen. The man's aged face shows what the girls' smooth skin cannot; it allows us to perceive movement around the eyes. The wrinkled face reveals nuances in the expression that, in a child, remain hidden from view. In exploring what prompts smiles and laughter, Darwin explained the facial movement and corresponding sounds as the result of stimulation of the nervous system. "Under a transport of Joy or of vivid Pleasure," he says in the third chapter, "there is a strong tendency to various purposeless movements, and to the utterance of various sounds." The less control the happy person has over his feelings, the more obvious the action of the nervous system becomes. In keeping with his belief that animals, children, and the insane uninhibitedly express their emotions, Darwin demonstrated the range of such "purposeless movements" with three examples: children clap their hands and jump up and down; horses and dogs leap about unrestrainedly; and, upon receiving a telegram, an inmate of the Wakefield mental institution went for a walk, came back singing, and finally threw up in excitement.[46] Darwin classified laughing as part of this repertoire of involuntary actions. He considered smiling to be a lesser form of laughter—so the girls' faces represent the nervous system derailing in its most charming way.

The reader of *Expression of Emotions* encounters Duchenne's old man three more times, all in the twelfth chapter, "Surprise, Astonishment, Fear, Horror." These illustrations show him in a state of "terror," in "great mental distress," and in the grip of "horror and agony" (Figure 62). While the second illustration reproduces Duchenne's photograph, the other two are based on photos but take the form of woodcuts. A comparison of the photographs and the woodcuts reveals that Darwin again made changes—this time more far-reaching ones. While the passepartout around the picture of the "false smile" in Plate III hid the bulk of the laboratory setup, the woodcuts eliminate all

Fig. 21. Entsetzen und Todesangst, copirt nach einer Photographie von Dr. Duchenne.

Figure 62. Woodcut showing "horror and agony" based on Duchenne's photograph in *Expression of Emotions*

signs of it: both the electrodes and the experimenter's hands visible in Duchenne's image are not present. Darwin's instructions on this matter survive. "Omit galvanic instruments and hands of operator," he wrote in a note to the engraver, James Cooper.[47] The same instruction was also followed for the other woodcut. All the images in which Duchenne's electrical instruments were visible were thus reworked by Darwin for his publication.

As Prodger discovered, the changes to Duchenne's photographs are not the only examples of retouching in *Expression of Emo-*

Figure 63. "Ginx's Baby," figure 1 at the upper right, from Plate I in *Expression of Emotions*, in the chapter "Suffering and Weeping"

tions.[48] Darwin's Figure 1 on Plate I, which seems to be the photo of a crying baby, is, strictly speaking, not a photograph at all (Figure 63). The image—a screaming infant sitting on a chair, its head turned to the camera and its arm bent—is actually a photograph of a drawing of a photograph. Both the drawing and the photograph are by Rejlander. A comparison of the original photograph and the photographed drawing reveals that the baby's expression has not been changed but simply emphasized. Prodger therefore suggests that the photo was likely reworked to avoid a loss of detail during the technical process of preparing it for publication.[49]

However, Rejlander not only intensified the child's expression—he also altered the setting. Instead of sitting on the floor as it was in the original photograph, the baby is now positioned in an easy chair—a change of location that makes the picture seem more intimate. Because the seat is child-sized, the infant also seems disproportionately large. Compared to the original, the now gigantic child sits sideways on the chair with its eyes squeezed shut, its hair curling and its incisors gleaming, framed by an oval and surrounded by the five other crying children on Plate I. Darwin would have known about the manipulation of the baby picture just as he would have known about the changes to Duchenne's old man, since the portfolios in the archive contain both the original and retouched versions of these examples side by side. Sometimes the alterations help hide elements of the picture, and sometimes they introduce new ones—for example, electrodes vanish, and an armchair appears. But whatever form they take, the changes always they serve to enhance the private, intimate atmosphere.

The debate about scientific photography emerging at this time shows just how traditional Darwin's use of the medium was. For example, Thomas Henry Huxley, now president of the Royal Ethnological Society, had already called for uniform standards for scientific photography in the 1860s so that pictures could be compared and data could be extracted from them—the "precise information respecting the proportions and the conformation of the body which [is of paramount] worth to the ethnologist." Darwin's photographs, in contrast, are marked by their very lack of standards. If we compare them to the photos of other scientists, such as Louis Agassiz's daguerreotypes, this difference jumps out.[50] In the 1850s Agassiz had commissioned a series of images of enslaved plantation workers in South Carolina with the purpose of creating a racial typology and demonstrating the superiority of what he considered to be the "white race." Agassiz was trying to replicate the distinction among types at the center of his frontispiece to his 1848 *Principles of Zoology* (see Figure 28) for humans in the same time frame. To distill these types, he excluded all personal details when shooting his models. The composition is identical from picture to picture: the models are naked and were apparently asked to keep their faces as motionless as possible; they always stand in the same pose; and the location is always an indeterminate interior space. This uniformity, identical position and framing, and lack of all personal details would later become the standard for scientific, ethnological, and anthropological photography. The approach represented the antithesis of portrait photography, which used location, accessories, expressions, and clothing precisely to communicate as much as possible about the subject.[51]

Darwin, on the other hand, never depersonalized the subjects of his pictures. Indeed, the very selection of photographs he made for *Expression of Emotions* documents his preference for shots of individuals rich in accessories, as most of the book's illustrations are from the studio photographers rather than from the physiologist Duchenne or the physician Crichton-Browne. Darwin chose eight illustrations by Duchenne and just one by Crichton-Browne, but twenty by Rejlander and another five from the Wallich and Kindermann photos. In contrast, the original archive contains the eighty-four plates in *Mécanisme*, as well as seventy-three photos by Rejlander, thirty-seven insane asylum photos by Crichton Brown, and sixteen images from the Kindermann photo studio in Hamburg.

This decision had consequences for the mood that is evoked in the photographs within *Expression of Emotions:* the world Darwin presented for his readers in engravings and photographs was not that of the laboratory, with its medical observations, experiments, and instruments, but rather that of bourgeois Victorian England's living rooms and portraits. The young lady, the bearded Rejlander, the smiling girl in the hat, and the baby on the armchair overlap seamlessly with the casts of Victorian novels, studio photographs, and tableaux vivantes. In this light it is not difficult to understand why Darwin found electrodes and researchers disturbing but accepted the presence of hats, chairs, flounces, and velvet jackets. Thanks to the pictures, his theory that expressions were physical reactions to physiological stimuli—and that these reactions were vestiges of an animal past—seemed less offensive. Indeed, they directly countered the accusations of his critics. The bestialization of humanity, so often cited as the necessary consequence of the theory of evolution, is nowhere to be seen. The photographs in *Expression of Emotions* leave the bourgeois self-image intact to an almost surprising degree—a fact that helps explain the book's initial success. The first edition of nine thousand copies sold out in just a few months, but sales of the second edition slowed considerably. This version was still available in stores after Darwin's death in 1882.

Later behavioral scientists have criticized the work as insufficiently scientific—a view that underscores Darwin's position as an exception to, rather than pioneer in, the tradition of scientific photography as later generations defined it.[52] In the early 1870s the direction scientific photography would take was still unclear, but in retrospect we know that the subsequent history of the portrait would be written elsewhere: in the fields of personal, artistic, or political photography. The individual—always at the center of Darwin's research—faded from the scientific world for decades.[53] In the end the greatest success fell to Rejlander. His picture of the crying infant in the armchair was soon dubbed "Ginx's Baby," after a popular novel by Edward Jenkins, and sold more than three hundred thousand copies,[54] making it the best-selling illustration from *Expression of Emotions*.

ANIMALS IN DRAWINGS

The other popular technique that Darwin used alongside photography to show the expression of emotions to his readers was the woodcut. Photography and woodcut, like human and animal, meet in the research material he collected just as they do in the pages of *Expression of Emotions* itself. In addition to the thirty-two photographs, Darwin included twenty-one woodcuts. The humans appear primarily in the former, while the animals are shown exclusively in the latter. It is in the woodcuts that the controversial aspect of Darwin's theory becomes apparent. The claim made in the photos that two muscle groups were involved in a human laugh was not likely to provoke resistance, but the woodcut's suggestion that animals also laugh—and use the same muscles as humans to do so—certainly was. Long exposure times made it impossible to photograph living animals well into the 1880s.[55] As a result, Rejlander, his wife, the old man, and the laughing girls take up positions among drawings of chickens and swans, dogs and cats, and chimpanzees and macaques. Furthermore, humans and animals were separated not just by the media in which they were depicted, but also by the order in which they were presented. First the animals, from dog to ape, take the stage, followed by the humans, from an infant to an old man. Contrary to what one might expect, the sequence from animal to mankind does not form an evolutionary progression from, for example, simpler mammals up to humans. The only part of the book structured on an evolutionary template is the point of transition between these two groups. The pouting ape that closes the discussion of animals is followed by the plate with images of several crying babies, whose expressions in Darwin's view more closely resemble the uninhibited animalistic demonstrations of their feelings than those of human adults, regulated by convention and upbringing. In this regard, ape and child are thus closer than ape and adult. Otherwise, neither the photographed humans nor the woodcut animals reveals an arrangement motivated by evolutionary theory. The humans represent various ages in no particular order, and the animals form a mix of taxonomic classifications. The first animal the reader of *Expression* encounters is a dog. More dogs follow, then cats, two spines from a porcupine, a hen, a swan, another dog and cat, and finally a macaque and a chimpanzee.

Like the photographs, none of the woodcuts was the work of Darwin. And, unlike the evolutionary diagram or pictures discussed in the previous chapters, no preliminary drawings or sketches survive. His contribution came in the form of tireless correction. The drawings, which he commissioned from artists or professional illustrators such as Briton Rivière, Thomas William Wood, and Joseph Wolf to serve as the basis for the woodcuts, bear his numerous comments. "The hairs on the neck and shoulders (and not on the loins) ought to stand closer (a series mass) and to be more erect," Darwin noted about a drawing of a bristling dog.[56] "Attend to ears in both figures," he admonished the engraver Cooper, who was transferring the drawing of a macaque to a wood block, "laid back & the wrinkles near eyes."[57] In his way, Darwin directed what he did not draw himself.

Almost all the artists with whom Darwin worked specialized in popular genres. Some were very successful, and from today's perspective some of their work borders on kitsch. Speaking of art, Darwin says at the beginning of *Expression of Emotions* that he has "looked at photographs and engravings of many well-known works," but adds that, in most cases, he has "not thus profited." The reason, he explains, is "that in works of art, beauty is the chief object; and strongly contracted facial muscles destroy beauty."[58] This observation holds as long as we consider the higher genres or, as Darwin says, "great masters." For the generals, prophets, kings, or the other worthy figures that populate large-scale historical paintings, grimacing or even simply smiling was indeed inconceivable. But this problem did not arise in art with less exalted subjects. Lower categories of art, such as animal or genre painting, were free to run the gamut from outré, curious, or bizarre to droll, precious, or touching—and thus to show all the attitudes of human or animal faces that interested Darwin.

In this vein, a sculpture depicting the most fantastic of Shakespeare's characters led Darwin to discover the so-called Woolnerian tip. The artwork in question, by the Pre-Raphaelite artist Thomas Woolner, shows a goblin-like Puck from *A Midsummer Night's Dream* (Figure 64). A kink in Puck's ear aroused Darwin's interest. As he composed his preparatory drawings for the piece, Woolner had come across several individuals with

Figure 64. Woolner's 1847 statue *Puck*, which depicts the character from Shakespeare's *A Midsummer Night's Dream*. Darwin's *Descent of Man* includes a drawing of Puck's ear. (University Library Cambridge with the permission of the Cambridge University Library Syndicate)

misshapen ears and determined through comparative study that this characteristic also occurs in apes. Darwin, who had sat with the sculptor for a portrait bust in 1867, saw this malformation as the "vestiges of the tips of formerly erect and pointed ears."[59] He included Puck's ear in *Descent of Man*, thanking Woolner for the drawing and model of the outer ear.

For the woodcuts in *Expression of Emotions,* Darwin worked with animal artists in London. The fact that he commissioned local artists to prepare these drawings points to an unusual quality all the animal subjects share: each is a zoo animal or a pet that lived in the capital and therefore could be observed and drawn there. Unlike ethnologists in the twentieth century, Darwin primarily studied domestic or captive animals, not those in the wild. This circumstance had wide-ranging consequences for the type of behavior he observed. Because of the proximity between humans and zoo animals and pets, the pictures typically do not show members of the same species communicating among themselves. Instead, humans are always involved: the cat rubs against its owner's leg, the dog grovels before its master, the ape giggles as a keeper scratches the soles of its feet, the chimpanzee pouts when it doesn't get an orange from its caretaker.[60] The pictures thus show a little-studied relationship that developed between humans and animals in the nineteenth century and that formed the background of Darwin's studies. Never before could so many animals be found in bourgeois living rooms—ensconced on the sofas, easy chairs, side tables, and carpets. Within a decade, the middle class had responded to the displacement, industrialization, and urbanization that had otherwise banished them to a world apart from animals, by populating their homes with fish in aquariums, parakeets in cages, and cats and dogs in baskets. Indeed, some groups of humans and animals drew closer together than they ever had before: the bourgeoisie and their pets.[61]

Obviously, the fate of pets is not representative of how animals in general were treated in the nineteenth century. During a time in which mass animal husbandry developed and slaughterhouses sprung up at the edges of cities—culminating in the egg-laying batteries, fattening factories, animal transport, and processing plants of the twentieth century—pets had an exceptional and protected position, forming a kind of privileged caste. Beasts of burden, cattle, and other animals used and consumed by humans do not make an appearance in Darwin's theoretical studies of animal behavior, and wild animals in their natural environments are similarly ignored. His attention focuses almost exclusively on domesticated animals, a perspective that he shared with the majority of the Victorian middle class. The pictures in *Expression of Emotions* show these animals in a new light: they serve not just as a kind of ersatz nature or exotic showpieces for the Victorian bourgeoisie, but also as a means to understand the proximity of humans and animals to one another.[62]

The widespread enthusiasm Darwin's contemporaries felt for zoo and circus animals and pets often took the form of amazement at their similarity to humans. "Dear Mr. Darwin," wrote Thomas Henry Huxley's son Harry after visiting the circus with Lady Lyell, "we have just come home from the circus . . . there was a nice horse whose name was D[illegible] . . . and at the last there were some monkeys riding on ponies."[63] Equestrian apes are just one example of the types of animals that delighted a broad audience with their all-too-human abilities. The tricks performed by a small mixed-breed terrier named Toby earned a full-page illustration in the *Illustrated London News*, as did Jerry, a mandrill "which drank grog, smoked a short clay pipe, and on one occasion dined (off hashed venison) with George IV at Windsor Castle." No new animal could arrive at the London zoo without a thorough introduction, in words and pictures, to the city's newspaper readers.[64] Between engravings of events from the war, monuments, politicians, or new technical achievements, the major illustrated papers showcased the zoo's new bird of paradise, the African antelope, the Australian platypus, the walrus, the hippo, the desert fox, and the tortoise. When the *Macacus lasiotus*, a macaque from China, arrived in England in 1868, the *Illustrated London News* devoted a full-page engraving to the animal and took even the effusive reporting of the day to new levels (Figure 65). Titled "A new arrival at the Zoological Society's Gardens, Regent's Park," the illustration shows the moment—dramatically condensed—when the newcomer meets the other apes in the enclosure. They, in turn, curiously inspect

Figure 65. Newspaper illustration, "A New Arrival at the Zoological Society's Garden, Regent's Park," from an 1868 issue of the *London Illustrated News*

the fresh arrival. The faces of cage's inhabitants reveal fear, amazement, and even horror, and their expressions and gestures are exaggerated almost to the point of caricature. The viewer is treated to a feast of gaping maws, bared teeth, bulging eyes, and ears strained erect.

Images like this one strive not just to depict the outer appearance of the animals as accurately as possible, but to represent their essence, character, and personality. This interest in animals as individuals persists throughout the nineteenth century. Chapter 1 discussed George IV, who commissioned a portrait of his giraffe in 1827, and Queen Victoria, who had her dogs Islay and Tilco, her macaw, and the parakeets immortalized in a group painting (see Figure 4). Eventually newspapers printed the portraits of talented animals like Toby and Jerry or documented the mood in the ape pen as a newcomer arrived. Animals became worthy subjects of portraits, and the Victorian well-to-do loved seeing them depicted as humans. And just as the worthiness of members of the bourgeoisie to be the subjects of portraits went hand in hand with their social advancement, the status of the portrait-worthy animal was also evolving. This new way of representing animals reflects and communicates the rights that at least that some of them were granted at this time.[65] In 1820 the Royal Society for the Prevention of Cruelty to Animals (RSPCA) met for the first time, two years later Parliament passed the first animal-protection law, and in 1860 Battersea Dog's Home, London's first animal shelter, was founded. The vegetarian movement and major antivivisection campaigns followed, culminating in 1876 with the Cruelty to Animals Act—a law that, while not forbidding vivisection, at least regulated and controlled it. In the 1860s Darwin himself and his wife, Emma, campaigned against the use of steel traps in hunting areas and reported a neighboring farmer to the RSPCA when he put horses to work in the fields when their necks were rubbed raw.[66]

The growing demand for animal pictures for books, newspapers, or private use led to the rise of a profession whose members worked closely with Darwin: the animal painters and illustrators who plied their trade in the zoos. No standard education or training was necessary to become an animal artist. Some members of the profession studied at art academies, others learned their trade in engraving workshops, and still others were self-taught. What they all had in common was their study of living animals. They took up positions with their sketchbooks and easels in front of the cages and enclosures at the Zoological Gardens to capture animals in their characteristic attitudes—their scampering movements or fleeting expressions. Some of them also kept animals such as birds in their studios or borrowed the smaller of the zoo's new arrivals to paint them in the peace of their own premises. The artists also frequented the dissections of alcohol-preserved zoo animals or other specimens that were contributed to the natural history museums. Like the generations before them, they also worked from other pictures or stuffed specimens. In some cases, they continued to propagate existing iconography, while in others—as we will see—they broke new ground.[67]

Even before Darwin, the work of animal artists created a tight bond between scientific institutions, such as the zoological society or the natural history museum, and popular ones, such as daily papers and publishing houses. Their work appeared in scholarly books and journals, then meandered into travel stories or children's books or turned up in the popular daily press. In some cases a single artist simultaneously supplied images to scholars, travel writers, and authors of children's books, while in others the illustrations found their way from publication to publication through the hands of a series of copyists. The gorilla, for example, whose career in the illustrated book market is examined below, made his first appearance in a scientific journal, then in the frontispiece of a travel book, and finally in cartoons and children's book illustrations. A brief overview of the activities of Darwin's artistic collaborators shows how varied the work of this group could be. Thomas William Wood, who drew a number of dogs and cats for Darwin, kept the general public informed of the new arrivals in the Zoological Garden in the *Illustrated London News*, edified the scripturally minded nature lover with drawings for Reverend Wood's *Bible Animals*, and illustrated animals for *The Field*, including the scene of the argus pheasants' mating dance (see Figure 52).[68] Joseph Wolf, who also worked with Darwin, produced numerous lithographs for the Zoological Society's *Proceedings* and *Transactions*, but his work also graced travel accounts, fables, and books for children. For *Expression of Emotions* he drew an ape, a horse, a cat, and a dog.[69] Wolf's outstanding ability to capture an animal's expression and movement also earned him the admiration of his artist colleagues. The Pre-Raphaelites "were delighted with his acute and minute observations, and delicate precision of rendering." Sir Edwin Landseer—the knighted portrait painter of the royal pets and creator of the lions for Trafalgar Square—supposedly cried out at the sight of a picture of a bird by his younger colleague, saying "he must have been a bird before he became human."[70]

The involvement of these popular animal artists meant that Darwin's scientific interests suited the public taste, and vice versa. Phenomena such as newspaper illustrations, the keeping of pets, visits to the zoo, bourgeois pastimes, and the involvement of laypersons in natural history intertwine within his research. And it was no coincidence that this research took place during a time in which animals came to be seen as worthy subjects of portraiture. Pets and zoo animals lived more closely with humans than with other members of their own species, and consequently they were given names and represented in portraits. The arrival of animals into bourgeois living spaces brought their individualization, and this closeness created a need to communicate across the border between species. This development is of decisive importance to Darwin's work: he was primarily interested in domesticated animal individuals and their ability to express themselves in signs and gestures comprehensible to— and similar to those of—humans. One example was Polly (Figure 66), the terrier belonging to his daughter Henrietta, which served as the model for the small dog the reader encounters in the first pages of *Expression of the Emotions*. Her head cocked to one side, ears perked, and left paw in the air, she attentively watches an object beyond the frame of the image, which the text describes as a cat on a table. The picture therefore portrays a domestic situation, although according to Darwin the lifted paw is the rudiment of a leap to capture prey—a reflexive movement that has become habitual. In a letter, Thomas Henry Huxley jokingly dubbed the animal that shared Darwin's study the "Ur-dog,"[71] thanks to her leading role in Darwin's behavioral studies.

The woodcut of Polly is based on a lost photograph by Rejlander, and the two dog illustrations that follow are the work of Briton Rivière, the most successful animal artist among Darwin's contributors. His career exemplifies the links that existed between evolutionary research and the animal artists at this time. Rivière took the stage during the years when Darwin began systematically collecting pictures of humans and animals. According to the artist's biographer, the painting that paved the way for his further success was *Sleeping Deerhound*, exhibited in 1865 in the Royal Academy.[72] Later Rivière would specialize in historical paintings featuring animals, including a Daniel

Figure 66. The dog Polly, the first animal to appear in *Expression of Emotions*

surrounded by six great cats in the lion's den, and Odysseus and his men as a herd of pigs in the painting *Circe*. Thanks to these efforts, he nearly became president of the Royal Academy.[73] In 1865, however, his subject was more modest: a sleeping English hunting dog whose head rests on its crossed front paws. Dog motifs had long been used in art, especially as allegories of faithfulness and loyalty. For example, numerous works since the sixteenth century had shown dogs at the deathbeds of their masters. In Rivière's picture no master was present, however. Freed from this traditional context to fill the canvas alone, the dog is seen from a perspective approaching that of animal psychology. The observer contemplating the sleeping dog does not think about what the animal means, but rather is left to wonder about its mental state. What is happening behind the animal's closed eyelids? Almost thirty years earlier, Darwin had posed the same question in his notebook: "A dog whilst dreaming, growling. & yelpings. & twitching paws which they only do when considerably excited, shows power of imagination—for it will not be allowed they can dream, & not have daydreams—think well over this;—it shows similarity in mind."[74] Both Rivière and Darwin had observed that sleeping dogs resemble dreaming humans. One man recorded this idea in a picture, the other in words.

In 1874 Rivière painted what his biographer considers the second key work of his career, two years after *Expression of*

Sympathy. From the picture in the Royal Holloway College collection. By permission of Messrs. T. Agnew and Sons, the proprietors of the copyright.

Figure 67. Engraving based on the 1874 painting *Sympathy*, by Briton Rivière

Emotions was published (Figure 67). *Sympathy* shows a small white dog and a young girl sitting together on the front steps of a house. The child has apparently been sent out and now sits out her punishment, her chin in her hand and her eyes staring up into space. The dog cuddles up to her, lays its head on her shoulder, and gazes into her face. As the title suggests, the dog's expressions and gestures convey sympathy and understanding. The large segment of the public that bought reproductions of the work apparently agreed. Darwin shared this view just as he had shared the opinion that dogs dream. "Besides love and sympathy," he wrote in *Descent of Man*, "animals exhibit other qualities connected with the social instincts, which in us would be called moral; and I agree with Agassiz that dogs possess something very like a conscience."[75] In *Expression of Emotions*, he calls the gestures and facial features of dogs and apes "almost as expressive as those of man."[76] When Rivière places the faces of the dog and child at the same level in *Sympathy* and thus encourages the observer to directly compare them, he also demonstrates this attitude.

Whether Rivière saw this similarity as proof of the kinship between humans and animals is not known—his views on the subject have not been preserved. His paintings exemplify observations shared by many pet owners, and not every pet owner believed in the theory of evolution. It is no accident, however, that the founder of evolutionary theory owned pets. Darwin repeatedly showed interest in accounts that suggested animals shared human qualities. When he first contacted Rivière in the summer of 1871, their correspondence focused on the ability of dogs to "grin" or "smile."[77] They questioned whether dogs lay their ears back or perk them up when they are happy, and how they express joyful excitement differently in response to food, prey, or the appearance of a trusted person. And they discussed Landseer's painting *Alexander and Diogenes* (Figure 68) from 1848, which represents the famous encounter in Cornith—using dogs. Alexander the Great takes the form of a white bulldog escorted by several bloodhounds, Cavalier King Charles spaniels, a pinscher, and a greyhound. Diogenes, in turn, appears as a bearded collie. Alexander the bulldog swaggers toward his counterpart with his chest thrust out and his head high, while Diogenes-as-collie lolls in his cask, friendly and relaxed. The picture may be difficult for modern viewers to swallow, but Darwin took it seriously. He kept a copy in his archive but eventually came to criticize Landseer's depiction of the animals' expressions.[78]

For *Expression of Emotions* Darwin commissioned Rivière to draw a dog with hostile body language and another in a state of happy excitement. The resulting drawings traveled four times back and forth between Rivière's home in London and Darwin's residence in Downe. Once Darwin asked the artist to indicate the

Figure 68. Landseer's 1848 painting *Alexander and Diogenes* (By the permission of Tate, London, 2010)

fur on the dog's neck more clearly; on another occasion Rivière resisted Darwin's suggested changes, insisting that dogs would not lower their heads at the sight of their master unless they were afraid. The dog in question in *Expression of Emotions* ultimately appears in the position Rivière described (Figure 69): Figure 6 of the book shows a pointer with laid-back ears and lowered shoulders that nonetheless stretches its head toward its master and meets his gaze. Describing the mouth of the excited animal, Darwin wrote, "The upper lip during the act of grinning is retracted, as in snarling, so that the canines are exposed, and the ears are drawn backwards." When it came to the frequency with which dogs grin, he deferred to Rivière's eye and experience: "Mr. Riviere, who has particularly attended to this expression, informs me that it is rarely displayed in a perfect manner, but is quite common in a lesser degree." Both men were apparently pleased by the successful collaboration. Instead of payment, Rivière requested a copy of *Origin of Species*. Darwin in turn referred to him in his introduction as a "distinguished artist" who had been so kind as to give him two drawings of dogs.[79]

Figure 69. Woodcut of a submissively affectionate dog, based on drawing by
Briton Rivière, in *Expression of Emotions*

DARWIN'S GORILLA

The dog and the ape are thus the two animals that most strongly
linked humans with the rest of evolutionary history, but in so
doing they served as counterparts to one another. Dogs bear
little physiological resemblance to humans but, according to
popular opinion, they share our spiritual and moral qualities;
apes, in turn, look more like humans but were seen as possessing no human virtues. In a nutshell, the anxiety provoked by the
idea that we might be related to apes was due to the fact that the
dog would have been a preferable next of kin. The iconography
surrounding both animals reflects this view. As it did in Rivière's
picture, the dog continues to feature in visual history as man's
best friend, and we take for granted the idea that dogs possess
advanced mental and character traits.[80] The horror that a mere
picture of an ape could evoke, in turn, can be illustrated by a
scene from Johann Wolfgang Goethe's *Elective Affinities*. In a
diary entry, Ottilie records her aversion after the first time she
leafs through a book with illustrations of apes. Ottilie, a representative of the eighteenth century, saw the ape as a caricature of
the human: "It is strange how men can have the heart to take such
pains with the pictures of those hideous monkeys. One lowers

one's-self sufficiently when one looks at them merely as animals, but it is really wicked to give way to the inclination to look for people whom we know behind such masks." The fact that apes were known to be intelligent and dexterous did not change the popular opinion of their characters. They were considered obscene, loud, and violent. At best, the animal could serve as a parody of the artist, in which the *tertium comparationis* between artist and animal was a relationship to reality based on mimicry and the "aping" of outer appearances. At worst, the ape could be found, as in Cologne Cathedral, among the spawn of hell. The iconography covered a signifying spectrum that ranged from ridicule to damnation.[81] Even the positive roles that apes had been given in literature since the Romantic period emphasized their outsider status rather than combat it. The ape takes its place in the gallery of half-men, half-beasts, including Victor Hugo's deformed Quasimodo and Mary Shelley's monster in *Frankenstein*. Of all the animals, the ape thus seemed the least qualified in moral terms to be humanity's relative, as a writer for the *Daily Telegraph* realized. "Even if Mr. Darwin and his friends could persuade us that our distant ancestors were guinea pigs or caterpillars," he reasoned, "people would not, we are inclined to think, found a new system of ethics on the discovery." The ape, in contrast, posed a moral challenge.[82]

The iconographic tradition affected evolutionary theory in sensitive ways. The fact that *Origin of Species* awakened interest in apes above all other animals—despite the fact that Darwin almost completely omitted mention of either apes or humans in the work—is well known. His precautionary measures could not prevent the ape from quickly becoming the mascot for his theory, nor the question of our relationship to the animal from being the central point in public discussion of it. In 1880, during a ceremony at Cambridge University to award Darwin an honorary doctorate, a wag lowered a stuffed monkey in academic regalia from the ceiling as the scientist entered the hall.[83] Even by the early 1860s evolutionary theory was known to the man in the street as the "monkey theory," in honor of the one aspect of Darwin's thinking that everyone could understand. If the various species had not been created by God, but had rather developed from a common ancestor, then—in this popular view—Adam must have been an ape. But exactly what kind of ape was still open for question. For the debate did not originate with an adorable monkey like the one lowered from the ceiling in Cambridge, but rather with a monster that haunted evolutionary theory like a shadow: the gorilla.

The pictorial history of the gorilla, which reached a temporary climax with the Du Chaillu picture shown at the beginning of this chapter, began surprisingly late, in 1852, when the first example of the animal reached Europe, preserved in a vat of alcohol. Earlier, in 1847, the Museum of Natural History in Boston had obtained a complete skeleton and scientists there named the creature *Troglodytes gorilla*, later *Gorilla gorilla*. The skeleton in Boston proved what researchers had long suspected: a third type of large anthropoid ape existed alongside the orangutan and chimpanzee, both of which had been known for some time. This large ape was seen by scientists for the first time in its complete form after it had been captured somewhere in the forest of West Africa, killed, preserved, and shipped to the Paris Museum of Natural History. The specimen put an end to the rumors that had circulated about the gorilla since 1625, when a travel author mentioned "two kind of monsters which are common in these woods, and very dangerous." This seventeenth-century account gives the larger of the two African monsters the name "pongo" and refers to the smaller as "engeco"—probably denoting the gorilla and the chimpanzee, respectively. According to the author, the pongo—the probable gorilla—is "in all proportion like man; but that he is more like a giant in stature than a man; for he is very tall, and hath a man's face, hollow-eyed, with long haire upon his browes. His face and eares are without haire, and his hands also. His bodie is full of haire, but not very thicke; and it is of a dunnish colour." The report adds that the pongo always walks upright on two legs, lives on fruit, is incapable of speech, and possesses "no understanding more than a beast." Furthermore, the pongos are said to be violent, forming packs, and killing "many negroes." The animals themselves "are never taken alive because they are so strong, that ten men cannot hold one of them."[84] Alongside its many frightening flourishes, the description contains information that identifies the forest monster as the gorilla: its vegetarian diet, color, size, and strength. But the diverse illustrations that such reports prompted over the centuries show creatures with little resemblance to real gorillas. Book illustrators were not traveling authors themselves, so for hundreds of years the gorillas were drawn by artists in Europe who based their depictions on mythological models.

The specimen that arrived at the Paris Museum of Natural History in 1852 settled many questions. Among them, it proved that this anthropoid ape from Africa was the world's largest primate, and that it was black from head to foot. However, we should entertain no illusions about the condition of an animal preserved in alcohol after an ocean voyage lasting several weeks. A few years later, when a similarly preserved gorilla arrived in London, an eyewitness reported that the stench of decay as the vat was opened would have been sufficient "to gas a company of soldiers."[85] The advanced state of decay of the Paris specimen led to several errors in the first picture of a real gorilla, which Mr. Bocourt, the museum's illustrator, published in the institution's in-house journal, *Archives du Muséum* (Figure 70 [Color Plate 15]).[86] Most obviously, Bocourt's gorilla is dark brown rather than black, a deviation explained by the bleaching effect of the alcohol. Second, the animal has a swollen belly, a sign of the bloating the body undergoes during decomposition. Third, the animal's teeth seem to slide out of its mouth almost like dentures, reflecting the fact that decay also causes the lips to become loose and rubbery. Despite these details, the animal that Bocourt drew is recognizable as a gorilla, although the matter-of-factness with which he portrayed his subject standing upright is remarkable. This upright being assumes an anomalous position between man and beast, and the branch on which the gorilla holds himself serves

Figure 70. The gorilla in 1858 *Archives du Muséum* from the Museum of Natural History in Paris (Color Plate 15)

as a sign of both his similarity and inferiority to humanity. The branch replaces the peeled stick that, in sixteenth-century iconography, was a standard attribute of the "wild man," an exemplary barbarian who could stand upright only by supporting himself on a stock.[87] The picture thus suggests that, like the wild man, the gorilla cannot stand upright on its own. Bocourt's illustration aroused the interest of the scientific community when it appeared in the museum's publication series in 1858, but there was little response from the general public.

This lack of general interest came to a sudden end in 1860 with the African journeys of Paul Belloni Du Chaillu, who, as we saw earlier, presented himself as the first Westerner to see gorillas firsthand. In America Du Chaillu's supposed encounters with live gorillas brought him immediate fame;[88] he went on lecture tours, exhibited his collection in Boston, and signed a book deal with the American publisher Harper and Brothers. In 1861 the Royal Geographical Society in London invited him to come to Europe and report on his West African travels. This invitation was largely thanks to England's colonial interests, since the country did not want to miss an opportunity to learn about the geography, raw materials, and natural resources such as rubber, dyes, and ivory in a region where French traders had long earned impressive profits. Furthermore, Richard Owen, the celebrated comparative anatomist and Darwin opponent, used his position as superintendent of the natural history collection to push for Du Chaillu's invitation and for the purchase of his specimens. The explorer addressed the Royal Geographical Society at the end of February, followed by appearances at the Royal Institution and the Ethnological Society. The sixteen gorilla specimens that he had brought back from his journey were treated and put on display, first in March 1861 in the rooms of the Royal Geographical Society in Whitehall Palace, and later in the gallery of the British Museum. The exhibitions occurred during the same year that Du Chaillu's book was released in England by John Murray, who had brought out *On the Origin of Species* two years earlier.

Figure 71. Illustrated plate, "The Shooting," from Du Chaillu's 1861 *Explorations and Adventures in Equatorial Africa*

Not surprisingly, Du Chaillu's views on evolutionary theory are not recorded; this debate probably was far afield from his interests. With the publication of *Explorations and Adventures in Equatorial Africa* in 1861, he stylized himself as a fearless hunter, adventurer, and eyewitness—as his book's famous illustrations show (Figure 71). In one example, a teeth-bearing, chest-beating, eye-rolling gorilla approaches two hunters in a forest clearing, barely a rifle's length away. The first is Du Chaillu himself, collected even in the face of danger and prepared to fire; the second is his African companion, who cowers behind him. The picture provided a model for many big-game hunters who followed. But Chaillu's success took a sudden turn. The initial salvo in a series of accusations came from John Edward Gray, head of the British Museum's zoological department and vice president of the Royal Zoological Society. Gray wanted to prevent the British Museum from buying Du Chaillu's African collection, which he considered too expensive and of questionable quality. To stop the deal Richard Owen had initiated, Gray started a campaign against the explorer. In the highly respected journal *Athenaeum* he accused Du Chaillu of plagiarism, unleashing the "Gorilla War" in the London press. "I may observe," Gray concluded after a cascade of accusations of duplicity and fraud, "that the frontispiece is copied from M. I. Geoffrey St.-Hilaire's figure of the Gorilla, published in the *Archives du Muséum* for 1858, from the specimen in the Paris museum."[89]

A comparison of the images proves that Gray was correct (see Figures 55 and 70). The two gorillas are identical to the last detail: the same pose, the same features, the same positioning of the head. One of the few differences that the copyist had permitted himself was to cover the animal's genitals, which were exposed in the original, with a branch. Otherwise the frontispiece was faithful to its French forerunner, including the anatomical errors that Mr. Bocourt made in the plate for the *Archives du Muséum*. Like its predecessor, Du Chaillu's gorilla had a distended belly and the same loosely hanging lips. Gray claimed that the picture had been drawn by Joseph Wolf and not, as Du Chaillu claimed, an American artist who had worked from the explorer's own sketches.[90] Inconsistencies in Du Chaillu's maps and the chronology of his journey, along with the patent falsity of his claims to be the discoverer of a number of West African species, all added weight to the accusation that the gorilla image was created in England and based on a French original rather than drawn in West Africa from life. The denunciations reached a climax in an essay in the *Proceedings of the Zoological Society* by William Winwood Reade, also an African explorer, which recounted in detail the reasons why the author believed Du Chaillu had never killed a gorilla himself or had even seen one alive. Reade was convinced Du Chaillu was a virtuoso counterfeiter rather than a witness: "M. Du Chaillu has written much of the gorilla which is true, but which is not new; and a little which is new, but which is very far from being true."[91]

Du Chaillu may have been exposed as unreliable in his accounts, but his pictures lived on. The date of their appearance lent their portrayal of the great apes a reach that neither Du

Figure 72. Caricature from *Punch*, 1861

Chaillu nor his opponents could control. The images became
their own argument. Intentionally or not, the debate about Charles
Darwin's evolutionary theory turned Du Chaillu's frontispiece
into a topsy-turvy vision of Paradise. Even *Punch*, England's
most popular satirical magazine, picked up on this implied competition with the biblical Adam when it portrayed the gorilla in
May 1861 (Figure 72). This infamous cartoon shows a gorilla
standing upright with a sign around its neck that displays the
question "Am I a man and a brother?" The words form a stark

contrast to the animal's appearance: the questioner is a savage, barbaric beast who, with a mouth full of razor-sharp teeth, pleads for admission into the human community. In the *Punch* caricature, a number of associations finally meet that had lain like a coordinating grid over the debate about our kinship to apes from the very beginning and lent this argument its intensity. "Monkeyana," the cartoon's title, is a reference to an identically named series of etchings from 1827 by Thomas Landseer, Edwin Landseer's brother. This successful work uses the animal in its traditional function as a distorting mirror, depicting a dressed-up ape to parody middle-class society.[92] But while Landseer's satire has the bourgeoisie as its target, the *Punch* cartoon plays on the debate about the abolition of slavery in a way a resident of Victorian England would immediately recognize.

The words "Am I not a man and brother?" first appeared in 1787 on a miniature produced by the porcelain manufacturer Josiah Wedgwood, Emma Darwin's grandfather. This ornament, which showed a kneeling slave in chains, served as a symbol of the abolitionist movement, and American opponents of slavery displayed the porcelain miniature on bracelets, barrettes, and snuff boxes well into the nineteenth century.[93] By replacing the kneeling slave with a begging ape, the *Punch* cartoon drew a parallel between the dark-skinned slave and the black primate. The fact that a joke comparing non-Europeans to apes could function thirty years after slavery was outlawed in England shows how insecure the status of such people remained. The cartoon thus feeds directly into the racial debates of the nineteenth century and the belief that some ethnic groups were closer than others to the animals. While the origins of this misconception had nothing to do with evolutionary theory, Darwin's ideas opened a door for it to reenter public debate. The racially motivated reduction of humans to animals is exemplified in an 1860 engraving from the *Illustrated Police News* (Figure 73). This drawing—which is not intended as a caricature—purports to document an incident in which a former slave in Jamaica dined on a child. The wild locks framing her head make it difficult to determine where her face

Figure 73. Illustration from the *Illustrated Police News*, 1861

ends and her hair begins. Her dress is filthy, and only her bared
teeth and protruding greedy eyes gleam through the layer of dirt.
She holds a small white leg in both hands, gnawing it like a
drumstick. A skull and other bones litter the floor of her disheveled hut. The difference she presents from the elegant white man
striding through the door could not be greater. The text informs
its readers that the woman ate twenty-six children before the
English authorities—embodied by the white man in the hat—put
a stop to this terrible behavior.

In retrospect, obviously almost everything about this scene has been invented, from the slave who eats children to the irreproachable servant of the English government. This realization provides an opportunity to recall the key points regarding England's involvement in slavery: the nineteenth century's greatest colonial power, England dominated the slave trade between 1730 and 1800 and imported the most slaves of any nation to its colonies. The country outlawed slavery in 1807, but not before bringing a final thirty-eight thousand slaves from Africa to the West Indies in 1806. The deadly conditions at the English sugar plantations caused mortality rates to reach 50 percent—in other words, one out of two Africans there died. Although two million human beings had been forcibly transported to the West Indian colonies during the 180-year duration of the slave trade, only 670,000 were still living there in 1834. Despite a steady supply, the slave population dwindled instead of growing, a situation that made slavery ultimately unprofitable. In fact, this economic argument contributed significantly to the elimination of the practice.[94]

The outlawing of slavery spelled the end neither of the widely held belief that non-Europeans were lower forms of humans, nor of depictions of Africans and Europeans as morally polar opposites. For centuries people have explained away their own bestiality by imagining others as monsters who deserve no better treatment or who are even dangerous. Alexander von Humboldt's famous statement that the idea that humans are descended from apes is "the greatest insult to mankind" since slavery makes sense when considered in this context. Humboldt, an opponent of slavery, knew that the slaves, not the ruling class, were the ones in danger of being written off as animals.

The fact that evolutionary theory postulated that all humans, including Europeans, shared a kinship to the apes did not eliminate racist attitudes. Instead, such beliefs resurfaced in the guise of evolutionary teachings in which the gorilla once again played a key role. While the gorilla-like appearance of the woman shown in the *Illustrated Police News* likely draws on older images in which slaves were portrayed with ape features, the debate about evolutionary theory lent such pictures a titillating new ambiguity.[95] It seemed to be just a matter of time before pictures such as Ernst Haeckel's infamous frontispiece to his 1868 *Natürliche Schöpfungsgeschichte* (Natural History of Creation) would make our kinship with the apes a question of race. In *Evidence as to Man's Place in Nature* (see Figure 46) from 1863, Thomas Henry Huxley had portrayed the gorilla as humanity's closest relative. Haeckel himself had also assigned the animal this role in two family trees (see Figures 34 and 35), and in his unpublished sketch Darwin agreed with them both (see Figure 54). Now, however, Haeckel was claiming that the gorilla was more closely related to some people than to others. His image (Figure 74) shows twelve heads in profile—six human heads presented above six ape heads. The series begins at the top left with a representative of the race that, according to Haeckel, was superior: the "Indo-Germanic," of which he of course considered himself a part. The form of ape he viewed as least developed, the baboon, is located at the bottom right. The heads between these extremes create a hierarchy of development from the most advanced human to the lowliest ape. The considerable distance between the Indo-Germanic peoples and the apes shrinks noticeably with the—in Haeckel's words—"African" and "Australian negroes," finally forming a hinge between the species at the point where the Tasmanians meet the gorillas. "The lowest humans . . . clearly are closer to the highest apes than they are to the highest humans," Haeckel remarked, a point of view that used evolutionary theory to give new life to old racist beliefs. Even Huxley included a table in his 1863 *Evidence as to Man's Place in Nature* showing several ape skulls along with that of a native Australian.[96]

At the same time, images of the gorilla seeped ever more deeply into popular culture, appearing in cartoons, magazines, children's books, and travel accounts. *Punch* featured a gorilla again in 1861, this time in a top hat and tails at a party, and dedicated

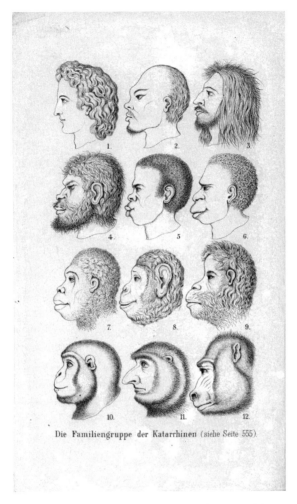

Die Familiengruppe der Katarrhinen (siehe Seite 555)

Figure 74. Haeckel's frontispiece for the 1874 *Natural History of Creation*
(Darwin Archive with the permission of the Cambridge University Library
Syndicate)

Figure 75. Illustration by Linley Sambourne for the children's book *The Water-Babies*, by Charles Kingsley

the New Year's Eve edition to the animal, who serves as Mr. Punch's alter ego and sits on his lap. Charles Kingsley's popular 1863 children's book *The Water-Babies*, a satire of evolution in the style of *Alice in Wonderland*, features the last representative of a nearly extinct species, who is shot by a hunter named Du Chaillu just as it howls and pounds its chest (Figure 75).[97] As part of the formula for a travel book's success, teeth-baring apes can appear in illustrations even if they are never mentioned in the text (Figure 76). The fact that Alfred Russel Wallace says not a single word about an orangutan attack in *The Malay Archipelago* did not stop the 1869 tome from enticing readers with just this bloody spectacle. On the frontispiece, five Bornean hunters have surrounded an unkempt orangutan, which flings itself on

ORANG-UTAN ATTACKED BY DYAKS.

Figure 76. Frontispiece to Wallace's 1869 *Malay Archipelago*

one of the men, seizes his shoulder in the claws of its right hand, immobilizes his spear with its left, and sinks its teeth into the hapless victim's chest. The human, armed with just a spear, seems defenseless before the teeth and claws of this beast—reason enough for one critic to doubt that the animal could share a common ancestor with humans. According to this line of reasoning, we would have lost the fight for survival. "Place a naked

Figure 77. Caricature from an 1871 issue of *Harper's Weekly*

high-ranking elder of the British Association in the presence of one of M. du Chaillu's gorillas," mused the author, "and behold how short and sharp will be the struggle."[98]

The debate about the gorilla, which had made the animal the most discussed anthropoid ape of the day, had raged for ten years by the time Darwin finally took a position on the kinship of humans and animals in 1871's *Descent of Man*. Shortly thereafter, the New York magazine *Harper's Weekly* imagined a personal encounter between Darwin and a gorilla (Figure 77). In a reversal of the usual perspective, the cartoon shows a gorilla reduced to tears because he feels insulted by the teachings of evolutionary theory. This time the animal bears its teeth not to attack, but in a mixture of disgust, complaint, and horror. Backed by the leaders of the "Society for the Prevention of Cruelty to Animals," the gorilla points accusingly at Darwin, declaring that the scientist has treated him unfairly. "This Man wants to claim my Pedigree," laments the gorilla. "He says he is one of my Descendents." "Now, Mr. Darwin, how could you insult him so?" challenges Mr. Bergh, the organization's founder.[99] The drawing thus turns the situation around and imagines what gorillas might think about having humans as their closest relatives. This reversal, while charming, misrepresents Darwin—he never claimed a privileged place for the gorilla among our animal relatives. In *Descent of Man* he relegates the animal to the sidelines, and in *Expression of Emotions* the next year—despite devoting long passages to the orangutan, chimpanzee, baboon, and macaque—he says nothing more about the gorilla than that it frightens its enemies with cries and that its "lower lip is said to be capable of great elongation."[100]

Darwin thus made a detour around the gorilla in his published work, despite having included it as our closest relative in the family tree he drew for his own use: nothing but vague suggestions about the gorilla's place in evolutionary history, not a word about Du Chaillu, and no mention of a debate that made headlines in the 1860s as the "Gorilla Wars." When the first living gorilla arrived in London's Zoological Garden, Darwin had been dead for years. But what he had heard about the animal during his life made him thoughtful. In the only passage in *Descent of Man* which discusses the animal as our potential relative, he speculates:

> In regard to bodily size or strength, we do not know whether man is descended from some comparatively small species, like the chimpanzee, or from one as powerful as the gorilla; and, therefore, we cannot say whether man has become larger and stronger, or smaller and weaker, in comparison with his progenitors. We should, however, bear in mind that an animal possessing great size, strength, and ferocity, and which, like the gorilla, could defend itself from all enemies, would probably, though not necessarily, have failed to become social; and this would most effectually have checked the acquirement by man of his higher mental qualities, such as sympathy and the love of his fellow-creatures. Hence it might have been an immense advantage to man to have sprung from some comparatively weak creature.[101]

It would be the last time that Darwin, who from this point would turn his attention exclusively to plants, insects, and earthworms, would bypass the debate being pressed upon him.

Earlier, in *Origin of Species,* he had remained silent about human evolution, and now he said nothing about the gorilla. But in 1872 Darwin filled this gap in his work by delivering a surprising punch line in *Expression of Emotions* that vividly demonstrates how he saw the relationship of humans and animals.[102] This last of Darwin's pictures to be discussed in this book was the result of a story that began in the spring of 1871, when an attendant at the Regent's Park Zoological Garden reported a remarkable occurrence. The *Cynopithecus niger,* a pitch-black crested macaque that had recently arrived in London from the Malaysian island of Celebes, could laugh (Figure 78 [Color Plate 16]). Resembling neither the high-pitched screams that emerged from the baboon cage nor the guttural, minor-keyed tones the chimpanzees produced from deep within their rib cages, the sound of the macaque from Celebes was a chuckle. In making a light, chattering sound it would pull back its small black ears, curl up the corners of its mouth, and bare its front teeth. The name of the animal attendant who was the first witness of this remarkable behavior was Mr. Sutton, who quickly informed Mr. Bartlett, the zoo director. He, in turn, lost no time contacting Darwin, the zoo's famous and regular visitor whose interest in animal behavior was well known. *Cynopithecus niger,* Darwin was assured, would laugh when petted, an activity it enjoyed.[103] Immediately after learning of the incident, Darwin asked the zoo director to recommend an artist to draw a portrait of the animal. He was directed to Joseph Wolf. "Mr. Wolf has got an eye like photographic paper," Bartlett assured the scientist. "It will seize on anything." Darwin reportedly then visited the artist in his London studio, a kind of aviary which Wolf shared with countless birds. During Darwin's visit, a finch that considered the premises to be its own territory attacked the intruding scholar, flying into his beard, cheeping and pulling on it until the distinguished guest burst into laughter.[104] This very amusement

Figure 78. The laughing macaque from Regent's Park in Darwin's 1872
Expression of Emotions (Darwin Archive [DAR 53.1], with the permission of the
Cambridge University Library Syndicate) (Color Plate 16)

demonstrated that a similarity existed between the human Darwin and the macaque: the researcher visiting the animal artist laughed just like the monkey that was to be his subject. Wolf was given the job.

In 1872's *Expression of the Emotions*, the *Cynopithecus niger* laughs in the first part of the book, followed by three small girls and the old man at the beginning of the second part. Darwin had Wolf create two depictions of the monkey's face. The same artist who, as Gray rightly guessed, gave Du Chaillu's gorilla its demonic grimace now portrayed an ape baring its teeth to completely opposite effect. The expression that had made the gorilla into a monster now demonstrates the macaque's kinship to mankind. The form of the drawing emphasizes this proximity: it is a bust portrait, a conventional way to portray individual humans. From one drawing to the next the animal lays back its shock of hair, pulls back its ears, slightly raises its eyebrows, and curls its lips as it opens its mouth to emit its laughing sound. The large black gorilla face with its threatening expression has been replaced by a small, cheerful countenance. In a reversal of the anthropoid ape gnashing its teeth, Darwin's gorilla, his little Adam, was a laughing macaque. The bridge between human and animal came not in the form of a bestial gorilla, but rather a monkey who could smile.

This monkey was a first in the field of natural history: no researcher had ever claimed to have seen a monkey laugh. And certainly no one had made a laughing monkey the centerpiece of a book. When Herbert Spencer, the sociologist and philosopher who provided Darwin with the expression "survival of the fittest," wrote his book *The Physiology of Laughter* in 1861, it never occurred to him that peals of laughter could come from any creature besides a human.[105] But in Darwin's book not only young girls laugh, but dogs and monkeys as well. In the text he adds that, according to Wallace, orangutans also laugh and that he himself has seen the same expression in chimpanzees. Another picture shows a chimpanzee from the zoo that, disappointed to not be given an orange, pushes its lips forward in a

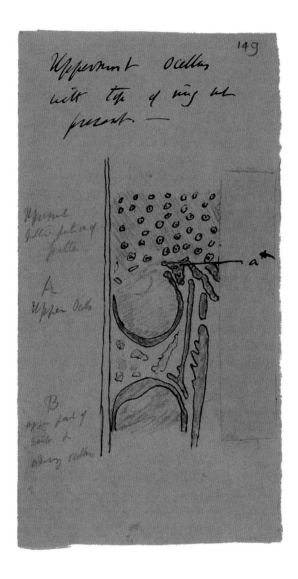

Plate 9. First of Darwin's three sketches of the argus pheasant's feather, circa 1867 (Darwin Archive [DAR 84.149], with the permission of the Cambridge University Library Syndicate) (Also Figure 37)

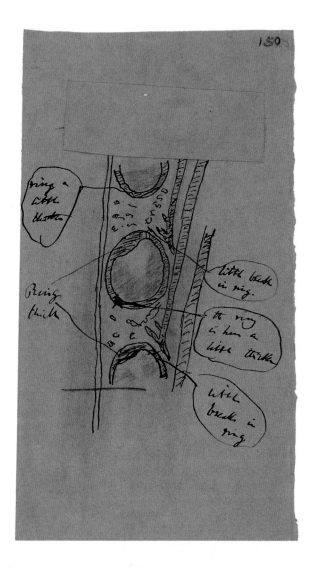

Plate 10. Second sketch of the argus pheasant's feather (Darwin
Archive [DAR 84.150], with the permission of the Cambridge
University Library Syndicate) (Also Figure 38)

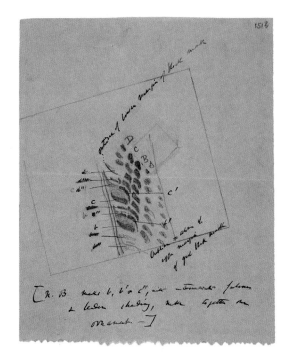

Plate 11a. Third sketch of the argus pheasant's feather (Darwin Archive [DAR 84.151b], with the permission of the Cambridge University Library Syndicate) (Also Figure 39)

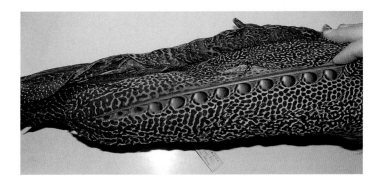

Plate 11b. Argus pheasant's wing from the former collection of the British Museum from the 1860s (Also Figure 40)

CHLOROSTILBON PORTMANNI.

Plate 12. Hummingbirds from Gould's *Monograph of Trochilidae*, 1849–1861 (With the kind permission of the Berlin Natural History Museum) (Also Figure 48)

ARTHUR LUCAS   49.WIGMORE ST.W

Plate 13. A stuffed argus pheasant in the form of a fireplace screen in a photo belonging to Darwin (Darwin Archive of the Down House Museum, English Heritage) (Also Figure 49)

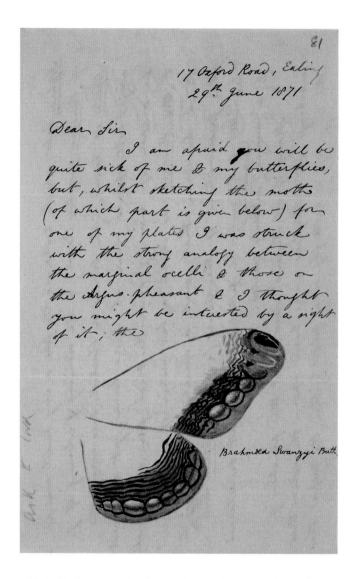

81

17 Oxford Road, Ealing
29th June 1871

Dear Sir

I am afraid you will be quite sick of me & my butterflies, but, whilst sketching the moth (of which part is given below) for one of my plates I was struck with the strong analogy between the marginal ocelli & those on the Argus-pheasant & I thought you might be interested by a sight of it; the

Brahmæa Swanzyi Butl.

Plate 14. Color sketch of an ocellus on the wings of a moth in a letter sent to Darwin (Darwin Archive [DAR 89.81], with the permission of the Cambridge University Library Syndicate) (Also Figure 50)

GORILLE GINA, GORILLA GINA MÂLE ADULTE
1/6¹ de grandeur naturelle.

Plate 15. The gorilla in 1858 *Archives du Muséum* from the Museum of Natural History in Paris (Also Figure 70)

Plate 16. The laughing macaque from Regent's Park in Darwin's
1872 *Expression of Emotions* (Darwin Archive [DAR 53.1], with
the permission of the Cambridge University Library Syndicate)
(Also Figure 78)

Figure 79. Frémiet's 1887 sculpture *Gorilla*

pout.[106] The readers of *Expression of the Emotions* thus held in their hands one of the best-natured books in the history of science. In its illustrations they saw laughing girls who could have been their daughters, the dogs and cats that shared their parlors, swans from Hyde Park, apes from the Regent's Park zoo, fussing babies, a young woman disappointed in love, and an eccentric London studio photographer. While Haeckel did not hesitate to depict Australians or Africans as the link between apes and humans, Darwin's visual argument rested instead on the observation that the English resembled their pets. His illustrations show neither wild animals nor non-European peoples. The pictures belie the claim that his most vehement critics used against him again and again: that being related to the animals would degrade humanity. For Darwin, this relationship dignified the animals instead. He demonstrated the link between animals and humans by showing not the animal aspects of humans, but rather the human aspects of animals. But humanity's fascination with the ape as a savage beast remained. The French sculptor Emmanuel Frémiet's famous work *Gorilla*, which won a prize at the Munich Art Exhibition in 1888, shows the animal carrying a white woman who struggles in vain (Figure 79). The path from Du Chaillu's gorilla to the twentieth century's *King Kong* was laid.[107] But the greatest opponent of this image of apes was the father of all monkey theories: Charles Darwin.

Conclusion

A BOOK about Darwin's pictures cannot end without mentioning what they do *not* show. Understanding the stages through which a picture passes between sketch and printed illustration allows us to experience Darwin as a strategist—a careful editor whose approach ranged from the fastidious to the tactical. He rigidly oversaw the preparation of all his illustrations, and, as the example of Duchenne's photographs show, his corrections could even take the form of retouching away original elements. Likewise, his books are full of scattered hints, such as the 1845 claim that "Light will be thrown on the origin of man and his history," that pay off only later. That he hesitated or put off making statements is one side of the story. Were there things he also left out completely?

Darwin entirely avoided anthropological illustrations in his published work. While numerous examples exist in his archive, any search for them in his published work would be in vain.[1] The *Journal of Researches* contains no images of Tierra del Fuegians, although he extensively discusses this South American people. In *Origin of Species* not a single person appears in either the text or the illustrations, and—with the single exception of a human embryo in the latter—neither *Variation Under Domestication* nor *Descent of Man* contains a picture of a human being.

People finally make an appearance in *Expression of Emotions*, but the individuals pictured are exclusively Europeans—primarily English, French, and German. A comparison with the work of his contemporaries, whether travel books or scientific studies, shows how unusual this is. Captain FitzRoy, on whose ship Darwin sailed around the globe, published his own account of the trip in the same year as Darwin and included several illustrated plates showing Tierra del Fuegians. And as we saw in the last chapter, the frontispiece in Haeckel's 1868 *Natürliche Schöpfungsgeschichte* (Natural History of Creation) featured humans of various races. Images of far-off ethnic groups and foreign-seeming people were standard fare in science and popular culture, from pictures in books to live shows.[2]

And Darwin? Darwin was an opponent of slavery and championed the view that all human beings share a common origin. He spoke against slavery several times in his travel book, and in 1846 he even asked John Murray if his criticism of the practice would be cut from the edition being published in New York.[3] As the historians Adrian Desmond and James Moore argued in their 2009 *Darwin's Sacred Cause*, the scientist's evolutionary work was fired by a moral passion. Without understanding his abolitionist convictions, they argue, "we cannot understand why Darwin came to evolution at all."[4] However, while slavery had apparently sensitized him about people of African origin, his comments on the Tierra del Fuegians are surprisingly derogatory and offensive. From the *Beagle* voyage he knew three such individuals: Jemmy Button, York Minster, and Fuegia Basket, a ten-year-old girl. A fourth member of their group had died on their way to England in 1830; now the three surviving members, after being raised in England, were returning to their homeland on the *Beagle*'s second trip. In *Descent of Man* Darwin counted the Fuegians among the "lowest barbarians"; in the same book he claims he would rather be descended from "that old baboon, who descending from the mountains, carried away in triumph his young comrade from a crowd of astonished dogs—as from a savage who delights to torture his enemies, offers up bloody sacrifices, practises infanticide without remorse, treats his wives like slaves, knows no decency, and is haunted by the grossest superstitions."

This curious comparison comes in part from Alfred Brehm's *Illustrirtes Thierleben* (Life of Animals), in which a story describes how one baboon saved the life of another. Referring to Brehm, Darwin retells the story in *Descent of Man*, contrasting the account with an episode from his travel book. There he had repeated the claim of another traveler in South America to have seen "a wretched mother pick up her bleeding dying infant-boy, whom her husband had mercilessly dashed on the stones for dropping a basket of sea-eggs!"[5] Darwin prefers the male ape to the cruel del Fuegian father as a relative; the South African native sinks here below the level of an animal. However, Darwin differs from scientists such as Haeckel in more than simply his appreciation of animals—disturbing as it may be at this particular spot—which marks all his work. Unlike Haeckel, he does not believe that Europeans possess physical characteristics that distinguish them from other peoples. Darwin indeed believed that Europeans were superior to other groups, but in his view it was not the head or skull of an Englishman that would make him a better person than the Fuegian, but rather it was the Englishman's upbringing. In 1879 Darwin sent a check to the missionary station in Ushuaia to pay for the education of Jemmy Button's grandchildren, but thanks to rebellions, epidemics, and war the Fuegian people vanished by the end of the nineteenth century. Writing about the disappearance of many ethnic groups during the course of colonialism, Darwin commented that "of the causes which lead to the victory of civilised nations, some are plain and simple, others complex and obscure."[6] That it was the English who had been cruel in invading the country in 1830 and kidnapping four of its people seems to have escaped him. But he still did not believe in a simple physiological or anatomical explanation for any group's victory or defeat. His strong belief in shared ancestry made a difference when it came to his book illustrations.

The archives of Cambridge and Down House contain bundles of anthropological illustrations from the *Collection anthropologique du Muséum* showing people from all parts of the world, as well as additional individual photographs.[7] In *Descent of Man* Darwin reports that when he showed some of these photographs to friends or family, they stressed the similarities among people the most. "This general similarity is well shewn by the French photographs in the Collection Anthropologique du Muséum de Paris of the men belonging to various races," he noted, "the greater number of which might pass for Europeans, as many persons to whom I have shewn them have remarked." Then he added a decisive observation: viewers are less prone to perceive this resemblance if the comparison does not involve anonymous photographs because we "are clearly much influenced in our judgment by the mere colour of the skin and hair, by slight differences in the features, and by expression."[8] Darwin, who reviewed every stroke in the drawing of the argus pheasant's ornament in *Descent of Man* and each line in his diagram in *Origin of Species*, knew that he would not be able to control the reception of ethnographic photos. Naturalists vastly disagreed about whether humanity consisted of a single group or many; scientists claimed there were from two to as many as sixty-three different species of humans. He thus concluded, as he wrote, that when it comes to races, "It is hardly possible to discover clear distinctive characters between them."[9] Since each individual believes he or she is seeing something different when looking at other human beings, the photographs would have been worthless as arguments for the unity of mankind. Darwin left the pictures where they still remain today—in the folders in his archive. His suspicions about the objectivity of photography have proven to be justified.

What Darwin likewise never—or almost never—shows is the act of selection or struggle. The only book illustration that touches on this subject actually shows an incident shortly before selection itself occurs: the argus pheasant bowing and unfurling the wheel of its wings (see Figure 52). Other nineteenth-century illustrations, in contrast, show struggle far more clearly: examples include Audubon's hunting eagle (see Figure 3), the attacking orangutan in Wallace's book (see Figure 76), or even Frémiet's gorilla carrying off a woman (see Figure 79). These images embody the eternal struggle in nature and the rule of the fittest—the very thing, in short, that became known as "Darwinism." Darwin quickly realized that it was possible to misunderstand the process of selection by compacting the concept into a fight between two organisms, a fight in which the physically superior one inevitably wins. To introduce what he means by "struggle for existence" in *Origin of Species* he thus uses the example of a plant that, along with other seedlings, struggles to survive at the edge of the desert. Those plants with a variation that allows them to store more water survive, while the others perish. In the sixth edition he adds another example—the flightless beetles of Madeira. The beetles with crippled wings survived, while those that could fly were carried off by the island winds to sink into the ocean.[10] In both examples, survival depends on adjusting to the environment, not winning a face-to-face fight. This interweaving of organism and environment is difficult to present visually. Nevertheless, selection became the subject of a veritable flood of images, especially in the twentieth century. From fighting dinosaurs to Neanderthals, they continue to make prehistory seem inseparable from battle and bloody struggle even today.[11] But the success of such violent visual motifs tells us more about the twentieth century than about Darwin's theory of evolution: whereas his followers concentrated on selection almost exclusively, Darwin's own pictures focus on the principle of variation.

Our walk though Darwin's visual archive has led through material of very different kinds. Pencil drawings and ink sketches are represented, as are prints, lithographs, woodcuts, and photographs. The relationship of the image to what it shows also varies widely. Some pictures, like the evolutionary diagram or the picture series, reveal processes invisible to the naked eye, while others depict concrete objects—finches, argus pheasants, or apes—and yet simultaneously illustrate concepts such as variation or kinship. Their functions were various: for the ornithologist Gould, pictures supplemented prepared specimens as a visual training ground for distinguishing among and classifying organisms. For disciplines such as geology, embryology, and, later, evolutionary biology, pictures provided a way to imagine periods of time we cannot experience directly. They brought the principles of evolution, variation, and selection together and showed the consequences of their interaction; they reconstructed the possible course of evolutionary history; they helped scientists to reduce ambiguity and record and communicate the things they observed—in the case of Gould's gold-leaf hummingbirds, for example, the perfection of nature and, in Darwin's countering image of the argus pheasant's ornament, its flaws. Pictures thus trained observational skills, made insight possible, developed theories, and communicated them to others. And again and again, they led a life of their own. Contrary to his intention, Agassiz's paleontological map of fossil fish became a model for illustrations of evolutionary theory (see Figure 27), and he was never able to fully reclaim it for his own purposes. Similarly, the teeth-gnashing gorilla was associated so frequently with Darwin's ideas that the animal came to serve as an image of the animalistic, archaic basis of humanity—even though the founder of evolutionary theory never suggested such a connection and countered this image with a picture of a laughing monkey.

Furthermore, the creation of Darwin's pictures was an enterprise necessitating the involvement of many people, and these images were facilitated by other pictures that had come before them. In the first chapter we saw how specialists and institutions scattered throughout London in the 1830s worked together, and the form this cooperation took.[12] In particular, the conclusions Darwin drew from Gould's effort to identify the *Beagle*'s bird collection show how a question of taxonomy classification could feed into the larger issue of whether species change.[13] While the first chapter thus reveals the interactions among the closely knit group of specialists centered in London, the following chapters evoke historical and geographic expansion. The strands of this larger network reached back in time and around the globe, from Downe in County Kent to the tropics of Malaysia. Hunting, transport, collecting, taxidermy, or zoos carried the objects of research over great distance and changed them at each step; geological maps, embryological atlases, anatomical plates, or illustrations of animals opened the conceptual framework for understanding an animal, an organ, or a plant arriving in England. As Darwin drew connections among these various approaches, what was formerly hidden became visible: the pattern of step-by-step change in the picture of the finches; the combined action of chance and selection in the knot of lines in the diagram or the picture series on the argus pheasant; the flaw in seeming perfection; the resemblance of human and animal. Finally, Darwin's pictures are indebted to the tradition of natural history, the elements of which they are composed. But they also break with these predecessors, piecing them together in surprising new ways. As a result, it is not a contradiction to claim both that Darwin's ideas are inconceivable without the naturalist tradition of the nineteenth century and that the theory that he developed was something quite new.[14] He embodied a seeming paradox: he was a revolutionary traditionalist.

All of these considerations relate to the relationship of images and science. While "image" and "art" are not synonymous, examining Darwin's pictures raises the question of how they respond to developments and issues in the arts. Ruskin and Millais admired Darwin's collaborator, the ornithologist John Gould; Strickland called the symmetrical presentation of the natural order "artificial" and its presentation as a map "natural"; Haeckel drew a distinction between collecting art and collecting nature; Darwin corresponded with animal artists such as Wolf or Rivière and commissioned animal studies from them, and he kept a copy of Edwin Landseer's *Alexander and Diogenes;* Woolner's statue of Puck served as the model for his

theory that a particular ear shape was a throwback; Bell criticized Rubens's lions in *Daniel in the Lions' Den;* Dickens compared Gould's hummingbirds with the works of Correggio; and
Darwin saw parallels between the argus pheasant and Raphael's
Madonnas, and between the process of creation and "chance
daubs of paint." The list goes on and on.[15]

A discussion of the relationship between art and science in
Darwin's work must distinguish between two separate levels.
The first level at which art and science meet is Darwin's collaboration with artists. The roles played by draftsmen, painters, and
sculptors is specified in the preceding chapters. Without Gould,
Swainson, Wolf, Rejlander, Rivière, Woolner, and perhaps Landseer, Darwin's work would have been inconceivable. They are
the invisible craftsmen in the history of evolutionary theory.
With the exception of Cameron and Rejlander, who have assumed places in photographic history, most of these Darwin
collaborators would not be represented in a book on nineteenth-
century art today. Some of Darwin's artists produced editions
in large numbers; their work was widely circulated, reached
broad classes of people, and significantly shaped the nineteenth-
century public's view of nature. Darwin preferred this kind of
popular art. However, he showed little interest in the art from
the period we most value now. He once traded an engraving by
John Flaxman for a billiard table, and Turner's watercolors left
him cold.[16] A group of illustrators and photographers joined
with him to write scientific history—but whether their efforts
represented art history is a different question. Those who have
seen Darwin exclusively as a revolutionary iconoclast may be
disturbed by this proximity to popular culture.

Art and science also meet at another level, in the form of the
question of the essence of nature. Darwin is commonly seen as
the founder of a functional conception of nature, but at the same
time the passages in his work expressing a Romantic view of
nature seem to form an arc to the perspective of art.[17] Both points
of view—Darwin as a materialist or as an artistically minded
aesthete—miss the point by presenting a false dichotomy. The

special thing about Darwin's view of nature is that it shows nature's flaws without denying its wonder. The argus pheasant was beautiful even as it was imperfect; the human eye is amazing despite being faulty. The closure of art faced the openness of a nature marked by the traces of its own becoming: chance, flaw, disorder. In finding the imperfect appealing or even beautiful, Darwin's concept of nature surprisingly resembles the modern aesthetic ideal.

Depending on one's viewpoint, the approximately hundred and fifty years between Darwin's theory of evolution and biology and the present can seem like either a long or a short period of time. Visual history links evolutionary theory back to the images and objects from which it emerged. The collections in London, the specimens in the British Museum, the studio photographers, the hummingbirds, the argus pheasants, and the apes in Regent's Park reinvest Darwin's theory with the local color that has peeled off during years of its reception, revision, and further development. In some cases, contemporary readers are likely to be familiar with the objects of Darwin's research, such as the Galápagos finches or fossils. Darwin's interest in the argus pheasant or household pets, on the other hand, is little known today.[18]

In the course of Darwin's reception, pictures have repeatedly shown a tendency to skip back to their earlier stages. Haeckel, for example, claimed to be following Darwin but repeated Chambers's 1844 diagram in his evolutionary family tree; the teeth-gnashing gorilla later seen as the embodiment of the evolutionary abyss came from an 1858 issue of a French museum's journal.[19] Faced with today's depictions of evolution, we can ask the same questions. What image of nature does an animal film present? Or a diorama of prehistory? Or a behavioral researcher? In some cases, the concept of nature they truly evoke may likewise be older than that on which they claim to be based. For example, Darwin's belief that female birds possess an aesthetic sense and choose their mates accordingly has been rejected. Instead, scientist support the theory that beauty is not an independent phenomenon, but rather a sign of health and strength.

According to current thinking, the female bird is genetically coded to recognize beauty as, for example, a resistance to parasites, and therefore to mate with the most striking male she can find.[20] This belief in the functionality of all animal characteristics seems to offer an uncanny echo of, of all things, Paley's natural theology.[21]

Finally, the visual history of the theory of evolution leads to the heart of Ernst Mayr's question of whether the life sciences are autonomous—in other words, if biology is a separate science from physics.[22] The tenacity with which Darwin's pictures deal with chance, variation, flaws, and individuals—the struggle for nonteleological representation they exhibit—proves that Mayr's view of the independence of biology was correct. In the twentieth century, Mayr faced criticism from physicist Ernest Rutherford, who accused biology of being "dirty science" compared with the mathematical disciplines. Similarly, Darwin's theory of evolution was dismissed by astronomer and physicist John Herschel as the "law of the higgledy-piggglety."[23] Following Darwin and Mayr, it is possible to see the idiosyncrasy of evolutionary biology as its strength. Equating biology and determinism, as so many recent debates do, is just as nonsensical in the light of Darwin's pictures as the essentialism that underlies these arguments.[24] "Nature versus nurture" discussions often suggest that variation, change, and diversity are due to the cultural factors in humanity's history, while the continuation and unity of features or behaviors represent our biological inheritance. Darwin's four icons proclaim the opposite. They also remind us that we are not at the end of evolution, but rather right in the middle of it. As long as evolution goes on, every organism it produces will be unique. Will the winds of Madeira favor the beetles with stumps for wings—or will the still air advance their brethren that can fly? In different forms, this question will be answered again and again.

NOTES

CUL: Cambridge University Library
DAR: Darwin Archive, Cambridge
DNB: Stephen, Leslie and Sidney Lee (eds.). 1993. *Dictionary of National Biography: From the Earliest Times to 1900.* 22 vols. Oxford: Oxford University Press.
Correspondence 1–15: Burkhardt, Frederick H., Sydney Smith, et al. (eds.). *1983–2005. The Correspondence of Charles Darwin.* Vols. 1–15 (1821–1866). Cambridge: Cambridge University Press.
Notebook A–N: Barret, Paul H., P. J. Gautrey, S. Herbert, D. Kohn, and S. Smith (eds.). 1987. *Charles Darwin's Notebooks, 1836–1844: Geology, Transmutation of Species, Metaphysical Inquiries.* London: British Museum (Natural History) and Cambridge University Press.

Introduction

1. *Descent of Man* appeared on February 24, 1871.
2. See *Fun*, July 22, 1871. For Darwin's many appearances in caricatures see Browne 2001. For the relationship of caricature and scientific history see Rudwick 1975 and Paradis 1997.
3. A review of Huxley's lecture can be found, for example, in the magazine *The Field* of April 16, 1870. Owen had already lectured on the horse series at the School of Mines in 1857. See Rupke 1994, 248–250. For the interpretation of Owen's horse series, see also Argyll 1867, 221. For the alternating prominence of Huxley and Owen thanks to their public appearances and the documentation of these events see White

2003, jacket illustration and Figure 3. For pictures as historical sources see Burke 2001.

4. The major biographies of Darwin offer excellent, detailed discussions of the extent to which Darwin's work and identity were conflated. See Browne 1995 and 2002, and Desmond and Moore 1991. For Darwin's celebrity in his day see Browne 2003 and Voss 2008. For the tactical maneuvers in this role-playing game, see Campbell 1989.

5. For its circulation in visual culture and society see Baumunk and Riess 1994 and Browne 2001. With the exception of Prodger's studies of the illustrations in *Expression of Emotions*, only isolated examinations of the pictures exist. See Prodger 1998a, b, and c and 2009; for more on the images of birds and plants see Smith 2006. The picture that has received the most study until now is Darwin's diagram; see Gruber 1988, Richards 1987, Richards 1992, Voss 2003a, Bredekamp 2005, and Brink-Roby 2009. For the new area of study of visual theory see Bachmann-Medick 2006.

6. See Browne 2001.

7. See Nochlin 2003; for the exhibition catalogues see Kort and Hollein 2009 and Donald and Munro 2009. See also Larson and Brauer 2009.

8. See Fritz Müller on Charles Darwin in an unpublished letter of January 16, 1872, in the Darwin Archive at Cambridge (abbreviated in the following as "DAR"), 142.55.

9. The archive of images that Darwin collected in preparation for *Expression of Emotions* has been systematically analyzed in Prodger 1998a. The archive at Down House also contains anthropological photographs. The expensive photographic illustrations for the printing of *Expression of the Emotions* led to Darwin's negotiations with, e.g., Murray in 1872; see Desmond and Moore 1991, 594, and Charles Darwin's letter to John Murray on July 27, 1845. The correspondence edited by Burkhardt and Smith and published from 1983 to 2005 is abbreviated in the following with title and volume, in this case, *Correspondence* 3, 230. The caricatures of Darwin can be found at DAR 140 and DAR 225.175–185.

10. See Gould 1995.

11. For the history of theories of evolution see Bowler 2003.

12. For the history of collections see te Heesen and Spary 2001, Beretta 2005, and Lange 2006; for its relation to Darwin's concept of a species see McOuat 1996 and 2001.

13. In *Darwin's Corals*, in contrast, Bredekamp interprets Darwin's diagrams as representational images of, not surprisingly, corals. To support this claim he overlays the diagram from the 1859 edition of

Origin of Species with an algae—which Darwin mistook for a coral—affixed to a page in a herbarium. Only one of the three arms of Darwin's evolutionary diagram is used in this comparison; furthermore, the herbarium page is reversed, two small arms are added, and another arm from the middle is eliminated (see Bredekamp 2005, 58 and 93). Such a fragmentary similarity does not support the far-reaching assumption that Darwin used this page as a model. The distinction Bredekamp makes between the nonhierarchically structured coral and the supposedly "hierarchically designed tree" (see Bredekamp 2005, 27) is likewise problematic. Haeckel first introduced such a hierarchical family tree in 1874 (see Figure 35 in this book); however, before this point, images of the "tree of life" took many forms, including numerous nonhierarchical ones (see Figure 22 in this book). The diversity of such tree diagrams meant that Darwin—like Wallace—could use the term "tree of life" throughout both his published and unpublished work while still creating diagrams of evolution that were not hierarchical.

14. For the preparation of animal specimens see Darwin 1958, 51, and Freeman 1978. Darwin wrote to Collier about the drawing class that "parts were too technical for me who could never draw a line [. . .]." See Darwin 1903, I, 398; for the *Beagle* notes see Darwin 1958, 77–78. For the history of drawing instruction see Kemp 1979.

15. For Haeckel's pictures see Gursch 1981, Krauße 1995, Kockerbeck 1997, Haeckel 1998, and Hopwood 2006. For Huxley's pictures see Bodmer 1997. For the close intertwining of Haeckel's life and work see Richards 2008.

16. See Darwin 1846. For Darwin's training in geology see Secord 1991, Herbert 1991a, Rhodes 1991, and Herbert 2005.

17. The original citation from Klee is: "Art does not convey the visible, but makes visible." Quoted in Burke 2001, 46.

18. For example, the English expression "struggle for life" was translated into German as "Kampf um's Dasein" and officially entered the famous German collection of citations *Geflügelte Wörter* ("Winged Words") in 1871. See Büchmann 1871, 84. Among works of literary theory, Beer 2000 offers a now-classic analysis. Also see Krasner 1992. For the dominant topos of struggle in the reception of the theory of evolution, see Ruse 1979 and Kleeberg 2004.

19. The importance Darwin granted to chance in his theory is a matter of debate. For example, according to the view championed quite early by Charles Sanders Peirce, Darwin considered chance to be a permanent, unavoidable factor in the life sciences. According to other interpretations, Darwin describes chance as a phenomenon that exists at the level of observation, and believed the study of inheritance would

eventually provide stable laws to explain it. See Peirce 1893 and Beatty 1984. The discussion of the role of chance in evolution is still taking place. See Reichholf 2003. When considering the ways in which chance is depicted in images, however, it makes little difference whether Darwin considered variations in species to be truly random or only appear as such.

20. The entire quote reads: "I am a complete millionaire in odd & curious little facts & I have been really astounded at my own industry whilst reading my chapters on Inheritance & Selection." See *Correspondence* 12, 337.

21. See Sternberger 1977, 79, also for the brilliant analysis of the ideologies that emerged from the theory of evolution. For the diversity of the worldviews derived from evolutionary theory see Rheinberger 2003b.

22. For the visual reception in the illustrated press in Germany and England see Voss 2009a.

23. In terms of the reception of the theory of evolution, such images preoccupied art in the nineteenth century. See www.19thc-artworldwide.org/spring_03/index.html. For Darwin's opposition to slavery and racism see Desmond and Moore 2009.

24. See Mandelstam 1991, 99f. My thanks to Peter Berz for drawing my attention to this passage.

Chapter 1. The Galápagos Finches

1. The other research ships were the H.M.S. *Sulphur* and the H.M.S. *Samarang*. See Browne 1995, 182.

2. For FitzRoy as the ship's second captain, see Browne 1995, 146; for the meeting of the BAAS, see Porter 1985, 988.

3. Humboldt's *Travels to the Equinoctial Regions of America in the Years 1799–1804* appeared from 1815 to 1832. The first edition of *Views of Nature* was issued in 1807.

4. For Humboldt's influence on later explorers and the attraction of the tropics, see Dettelbach 1996 and Stepan 2001. For Darwin and Humboldt see Kohn 1996. For Darwin's diary entry, see Keynes 1988, 42. For Darwin's letter, see *Correspondence* 1, 122, and Secord 1991a, 144f.

5. See Keynes 1988, 53.

6. See Lightman 1997. For the close links between biology and commerce that already existed in the eighteenth century, see Müller-Wille 1999.

7. For Charles Jamrach see Gunther 1975a, 269 f; for Linnaeus see Müller-Wille 1999. Stafford 1984 and Rice 1999 provide overviews of voyages of exploration.

8. In April 1831 Charles Darwin earned his baccalaureate in theology. At the recommendation of the Cambridge professor John Henslow he received the offer after Leonard Jenyns, a fellow Cambridge student, declined. See Browne 1995, 149ff.

9. In his autobiography, Darwin describes how he enthusiastically collected beetles as a child. See Darwin 1958, 22 and 62–63. Sloan stresses that Darwin had some knowledge of mollusks before setting off on his voyage. See Sloan 1985. However, his knowledge of other types of animals—including birds—was limited. See Steinheimer 2004, 3–5. For the onboard catalogs, see Sulloway 1982b, 332. For the practice of cataloging, see Swainson 1840, 6.

10. The travel account *Journal of Researches* appeared in 1839, the five volumes of *Zoology of the Voyage of the H.M.S. Beagle* from 1838 to 1843, and the geological studies *The Structure and Distribution of Coral Reefs, Geological Observations on the Volcanic Islands,* and *Geological Observations on South America* from 1842 to 1844. In addition to barnacles, the subclass Cirripedia includes certain types of balidinae.

11. In 1842 Darwin wrote an initial summary of this theory that he expanded into an essay in 1844. See Darwin and Wallace 1958. In his notebooks Darwin began speculating about evolution in 1837; see Chapter 2. Stott convincingly argues that Darwin delayed publication in order to confirm his ideas with further research. See Stott 2003, 135–153. See also van Wyhe 2007. Darwin received the Royal Medal in 1853.

12. See Darwin 1859, 488.

13. The name and classification of the Galápagos finches has changed several times: they were eventually renamed "Darwin's finches" and are taxonomically grouped with the New World buntings. See Steinheimer, Dickinson, and Walter 2006. Island populations in general have proven to be rewarding objects of study for scholars of evolution. From the perspective of evolutionary theory, the Galápagos islands are no more interesting than, for example, the Malay archipelago studied by Wallace, the Hawaiian islands, or Madagascar. The same is true of the finches. Other groups of animals could have equally well served as evolution's poster children. On adaptive variation on the Hawaiian Islands see Quammen 1996, 232f and 319f.

14. See Darwin 1845, 308.

15. Darwin's 1844 essay closes, for example, with a sentence about infinitesimal variation: "There is a grandeur in this view of life . . . that from so simple an origin, through the selection of infinitesimal varieties, endless forms most beautiful and most wonderful have been evolved." See Darwin and Wallace 1958, 254. For the pigeons see Secord 1981.

16. See Darwin 1872b, 107.

17. For the examples of its use in modern textbooks, see Steinheimer and Sudhaus 2006.

18. For Sulloway's groundbreaking reconstruction on which much of this chapter draws, see Sulloway 1982a, 1982b, and 1982c, and Steinheimer and Sudhaus 2006. Astoundingly, however, Sulloway considers neither the history of the picture's creation nor its reception. For more on the "invisible technicians" whose skills play an unseen but indispensable role in instigating discoveries and theories, see Shapin 1989.

19. For the history of the natural history collections of the British Museum see Gunther 1975a and Yanni 1999, 111–147. The Hunterian Museum originated with the collections of the physician John Hunter and opened in 1813. See Yanni 1999, 46–51. For the history and origin of the Zoological Society see Blunt 1976, 23–31, and see Desmond 1985 for a thorough account of the political background. The Zoological Garden in Regent's Park opened in April 1828 and attracted one hundred thirty thousand visitors in the first year. Zoo visitors who were not members of the Zoological Society were charged a shilling. See Desmond 1985, 228.

20. Darwin shared his reservations about various institutions and specialists with John Henslow in a letter. See *Correspondence* 1, 149, discussed in Porter 1985, 977.

21. On the allocation of Darwin's collection see Porter 1985. For FitzRoy's bird collection see Sulloway 1982c, 67–71.

22. George IV was not the first ruler to lose his heart to an exotic gift. Another famous example is the loving care lavished by Leo X on the white elephant Hanno, who arrived in Rome as a gift of the Portuguese king in 1515. See Bedini 1997. For the giraffe's arrival, see Tree 1991, 1–5, and Blunt 1976, 71–84. For J. L. Agasse's *The Nubian Giraffe, Its Native Keepers and Mr. Cross* and R. B. Davis's *Two Giraffes Belonging to George IV* see Blunt 1976, 76 and 177. Both pictures were painted in 1827 and are still owned by the British royal family.

23. The complete passage is: "midway between the eye of the finest Arab horse and the loveliest southern girl, with long and coal-black lashes, and the most exquisite beaming expression of tenderness and softness, united to a volcanic fire." See Tree 1991, 4. For the taxidermic preparation see Tree 1991, 6–12.

24. Publication of the *Proceedings* began in 1833 and the *Transactions* in 1835. Both journals still exist today.

25. On John James Audubon see Rhodes 2004; Blum 1993, 88–118; and Lambourne 1990, 78ff. On Darwin's reading of Audubon see Donald 2007, 81f, and Smith 2006, 104f. As Kemp points out, fitting life-

sized illustrations of the birds on a page required "twisted necks, bent legs, and curved bodies for the bigger species." See Kemp 2000, 53. On the history of ornithology see Stresemann 1951.

26. "A circle of animation and destruction goes perpetually round." See Smellie 1790, 398, and for the caterpillars, 396. For the passages underlined by Darwin see Di Gregorio and Gill 1990, 758. Darwin included Smellie in his list of reading as early as Notebook C. See Notebook C, 268. Martin Kemp comments: "Appealing and decorative as his designs may be, the untamed world of action he depicts is closer to Darwin's than to the standard bird book." See Kemp 2000, 53, Smith 2001, 58, and Smith 2006, 95f. Smith also examines the influence that Audubon's depiction of struggle and violence, especially that in *Birds of Britain,* had on Gould and further argues that it shaped Darwin's views of birds as well. See Smith 2001, 61ff. For the picture of the mockingbird see Plate 21, "Northern Mockingbird," in Audubon's *Birds of America,* reproduced in Blum 1993, 108 and 101. Darwin would later quote the botanist De Candolle on the "war of nature." See Notebook D, 135e.

27. Audubon came to England looking for a publisher in 1826 with about two hundred drawings in his bags. See Lambourne 1987b, 87. On the market for bird books see Tree 1991, 21.

28. The book described is by the American Daniel Giraud Elliott. See Palmer 1895, 112f.

29. For more on pheasants and the nobility, see Desmond 1985, 225. For the peacocks in parks see Wood 1862, 605. For Sir Edwin Landseer's *Islay, Tilco, a Macaw, and Two Love-Birds* see Lambourne 1990, 160f. For Joseph Wolf's portrait of the royal bullfinch see Palmer 1895, 223. For the hummingbird exhibition see Tree 1991, 171–177.

30. See Brontë 2002 [1847], 9. For Bewick see Uglow 2006.

31. See Ivins 1953, 86f., and Rudwick 1976. Bewick's new woodcut process combined various minor innovations: he began to cut the figures in the end instead of the horizontal grain; in other words, he used the surface that results when a trunk is sawed in horizontal sections rather than vertical ones. He also used a burin with a V-shaped tip rather than a knife. These improvements made it possible to portray fine distinctions of shading, raising the woodcut to the level of the copperplate engraving. Furthermore, the wood fibers proved to be remarkably resilient when pressed at a right angle to the grain rather than along it. Editions of one hundred thousand or more copies could be printed from a single block before wear became apparent.

32. See Anderson 1991 and Brake 2001.

33. In his overview of the history of ornithology, Stresemann calls Swainson one of "the most prolific and widely read ornithological writers in England." See Stresemann 1951, 177. Also see Farber 1985. Among William Swainson's books are 1835's *A Treatise on the Geography and Classification of Animals* and several volumes with the title *Zoological Illustrations,* published between 1820 and 1833, which introduced the technique of lithography into natural history in England. His dogmatic adherence to the classification system of William McLeay caused him to fall into discredit and he never received a post at a London institution. He immigrated to New Zealand, where he died in 1855. See Farber 1985.

34. On beauty as a subject of evolutionary theory, see Menninghaus 2003; also see Chapter 3.

35. The book on birds Darwin mentions by name in his autobiography is Gilbert White's 1789 *Natural History and Antiquities of Selborne.* See Darwin 1958, 45. For the visit to Audubon's lectures see Darwin 1958, 51. For his knowledge of ornithology see Steinheimer 2004. For his observation that children enjoy looking at pictures of animals, see Darwin's Notebook M, 28, quoted after Barrett et al. 1987. For his copy of Swainson's book see Di Gregorio and Gill 1990, 795ff. Noting the relationship of zoology and pictures, the editors of the index of artists working for the Zoological Society noted: "One could no more imagine the great zoological journals without their plates than a zoo without the animals." See Root and Johnson 1986a, xiii; and Root and Johnson 1986b.

36. On the origin of Gould's *A Century of Birds from the Himalaya Mountains* see Tree 1991, 27–33. All following information on the time and place of their meeting is taken from Sulloway 1982b.

37. See *Correspondence* 1, 525.

38. On the date of the meeting see Sulloway 1982b, 356. Darwin had already attended meetings of the Zoological Society in the autumn of 1836. Whether he met John Gould there is not known. See *Correspondence* 1, 513f.

39. John Gould's *The Birds of Europe* appeared in twenty-two installments, from 1832 to 1837; *A Monograph of the Ramphastidae, or Family of Toucans,* from 1833 to 1835; *A Monograph of the Trogonidae,* from 1836 to 1838. On Gould's life story see Dickens 1851. For Attenborough see the foreword to Lambourne 1987b, 7.

40. Alois Senefelder patented the lithographic technique in 1801, but almost two decades passed before interest in the craft developed. A translation of Senefelder's work appeared on the English book market in 1819 as *A Complete Course of Lithography,* and a second work by

Charles Joseph Hullmandel, *The Art of Drawing on Stone,* followed in 1824. For the frontispiece see Hullmandel 1824.

41. Little is known about Elizabeth Gould. Before her marriage she worked as a tutor for Latin, French, and music students. Whether she was trained in drawing—or perhaps even gave instruction in the field—is unknown. See Tree 1991, 23f. At the end of 1830, a year after their wedding, Gould offered his wife's drawings to the editors of *Illustrations of Ornithology,* a multivolume work that had been appearing in installments since 1825. See Tree 1991, 21, and Jackson and Lambourne 1990, 106ff. McEvey 1973, 11, emphasizes the importance of Gould's sketches for the quality of the illustrations in his books.

42. Research is unearthing the identities of more and more assistants from Gould's empire. For more on the artists, see Tree 1991, 34–48. On the role of colorists and the production process see Jackson and Lambourne 1990 and Lambourne 1987a, 50ff. For Gould's secrecy regarding his employees see Lambourne 1990, 191.

43. For these quotes, see Lambourne 1987a, 72; Lambourne 1987a, 67; and Jackson and Lambourne 1990, 196.

44. See Tree 1991, 214.

45. Gould could not have posed for the picture because Millais painted it after Gould's death. For more on the picture and on Ruskin see Tree 1991, 215f.

46. For the first two citations see Swainson 1840, 65. For the third, see Swainson 1840, 2.

47. See Swainson 1840, 1. For a detailed and insightful discussion of the connection between drawing and zoology in nineteenth-century America, see Blum 1993, especially 17–26.

48. For Gould's subscribers see Tree 1991, 164. For the practice of artists such as Antonio Pisanello, Albrecht Dürer, or Hans Baldung Grien of drawing based on stuffed bird specimens, see Schulze-Hagen, Steinheimer, Kinzelbach, and Gasser 2003.

49. For Audubon's relationship to Gould see Tree 1991, 18. In 1840 Gould had already commented: "We trust the author will hereafter reprint these expensive volumes in such a form as that they may be accessible to naturalists; and thereby diffuse science, instead of restricting it to those only who are wealthy." See Tree 1991, 125. For Ruskin see Ruskin 1906, 157.

50. Various existing taxidermic techniques for preparing bird specimens differed only slightly. This description follows the instructions provided by William Swainson. See Swainson 1840, 43ff. For the history of bird specimens and their significance for ornithology, see Schulze-Hagen, Steinheimer, Kinzelbach, and Gasser 2003, and Farber 1977.

51. Gould was named superintendent in 1833. By 1840 the bird collection was already considered superior to that of the British Museum. See Stresemann 1951, 144.

52. For Darwin's medical studies in Edinburgh see Browne 1995, 36–64. For his father's decision to have Darwin study theology after his medical education ended unsuccessfully, see Browne 1995, 89ff.

53. For Darwin's intention to become a clergyman see Browne 1995, 322–324. For his geological ambitions see his letter to Catherine Darwin of November 8, 1834, in Herbert 1991, 168, and Secord 1991a, 133. In November 1836 the Royal Geological Society elected him to membership, and in 1838 he was appointed the society's secretary for three years. See Rudwick 1982, 190. For "clergyman naturalist" see Moore 1985.

54. Quoted in Sulloway 1982b, 357.

55. Gould 1837, 4ff. The publication of the Zoological Society mentions fourteen species. In Darwin's works—both *Journal of Researches* and *Zoology of the H.M.S. Beagle*—this number is thirteen. Apparently Gould changed his classification once more in the course of his legwork for Darwin. See Darwin 1839, 461, and Darwin 1986 [1841], 98–105.

56. On the differences among the species of Galápagos finches see Steinheimer and Sudhaus 2006. On animal taxidermy see Rheinberger 2003c, 11f. In a breakdown of the various types of preserved specimens in medicine, zoology, and botany, Rheinberger defines the taxonomic specimen as used in zoology in this way: "On the one hand, [anatomical specimens] can show particularly typical examples of a class of objects. The specimen thus becomes a 'representative' of a class of objects, standing prominently for it and its characteristics." For the history of the species type in biology see Farber 1976. According to Farber the idea of a species type—in the sense of an object that allowed a species to be described for the first time—caught on in Europe's major natural history collections in the 1830s. Its acceptance was due to both the explosive growth in collections and new taxidermic techniques that made it possible to preserve animal cadavers. Also see Farber 1977. The case of a "type" for an organism is obviously much more complicated than that of the prototypical meter. Natural objects must exhibit "typological excessiveness" before they are granted the status of being a type. See Rheinberger 2003c, 12. For the paradox presented by species types, which stand in for a species without necessarily being typical, see Daston 2004.

57. See Gould 1837d, 27.

58. See John Gould's letter to William Jardine of January 1, 1837, quoted in Tree 1991, 51. Sulloway describes Gould as "notorious in his

efforts to obtain priority in the naming of new species." See Sulloway 1982b, 376. According to ornithologists today, the birds Darwin collected represented nine of the known species of Galápagos finch, not thirteen or fourteen as Gould claimed. See Sulloway 1982a, 40. However, taxonomists often disagree on how precisely a genus should be divided into species or subspecies. Champions of a nomenclature that allows grouping as many taxa as possible together and who emphasize connections are called "lumpers," while those who in support of making as many distinctions as possible are "splitters." Evolutionary biologists tend to favor "lumping," while taxonomic specialists—like Gould—show a preference for "splitting." See Campbell and Lack 1985, 330; also Mayr 1982 [1942], 282f.

59. Thank you to Frank Steinheimer for pointing this out in a discussion with me.

60. On the role of practical craftsman's knowledge in science, see Alpers 1983, 72f., and Rheinberger 1997.

61. Darwin relates the story in the account of his travels. See Darwin 1845, 92f. For more on taxidermic exercises see Swainson 1840, 50. For the practice of identification see Gould 1837e, 35. However, Gould failed to notice that the bird had already been identified. For this reason it is known today not as *Rhea darwinii*, but rather *Pterocnemia pennata pennata*, as d'Orbignys initially named it in 1834. See Steinheimer, Dickinson, and Walters 2006, 13.

62. "I feel sure as long as species-mongers have their vanity tickled [. . .]." See *Correspondence* 4, 187.

63. For the development of Darwin's theory in his notebooks, see Ospovat 1981, Sulloway 1982a and 1982b, and Gruber 1988. An overview of the starting points can be found in Oldroyd 1984 and, most recently, Hodge 2003.

64. The term *dinosaur* was later introduced by Richard Owen at a meeting of the British Association for the Advancement of Science in Plymouth in 1841. See Rudwick 1992, 141f.

65. Darwin dedicated the second edition of his travel book to Lyell, writing: "To Charles Lyell, esq., F.RS. this second edition is dedicated with grateful pleasure, as an acknowledgment that the chief part of whatever scientific merit this journal and the other works of the author may possess, has been derived from studying the well-known and admirable *Principles of Geology*." See Darwin 1845. For Darwin and Lyell see Herbert 1985. For the concept of species constancy see Lyell, "Whether Species Have a Real Existence in Nature," in Lyell 1990–1991 [1830–1833], ii, 29ff. Also see Rudwick 1992, 49. On criticism of the Great Flood theory in geology in the 1820s see Secord 1991a, 136.

66. Darwin's emphasis. See Darwin 1933, 383. Quoted in Sulloway 1982b, 351.

67. See Darwin 1845, 377.

68. See Sulloway 1982b, 348.

69. Later in his travel book Darwin wrote: "The tortoises differed from the different islands, and that he [the vice governor] could with certainty tell from which island any one was brought." See Darwin 1845, 394; for the tortoises and Darwin see Sulloway 1982b, 338–345.

70. See Darwin 1963 and Sulloway 1982c, 53. On determining the place and time at which Darwin catalogued his bird collection, see Sulloway 1982b, 334f.

71. Sulloway assumes that the meeting of Gould and Darwin took place between March 7 and 12. See Sulloway 1982b, 366.

72. The qualities of this paper are noticeably different from those of the paper Darwin used for the rest of his notes. The watermark and the small drawing of an animal suggest that it came from the Zoological Society and was provided to Darwin by Gould. See Sulloway 1982b, 363f.

73. Gould and Darwin presented a fourteenth finch together to a meeting of the Zoological Society on May 10. See Sulloway 1982b, 365. For the fact that Darwin's books specify only thirteen species, see note 55.

74. Quoted in Herbert 1980, 9. Also see Sulloway 1982c, 369.

75. For the fossils see Rachootin 1985 and Keynes 2003.

76. Darwin's notebooks have been transcribed and published by Barrett et al. 1987. All the following citations are taken from the transcriptions of the notebooks in this edition. The citations are identified by notebook and page. See Notebook B, 37.

77. See Sulloway 1982c.

78. See *Correspondence* 11, 405.

79. James Cook, who up to his murder on Hawaii had sailed around the world three times, visited Australia during his first world voyage in 1770. He reached the continent's southeastern coast from the Pacific Ocean, claimed the land for Great Britain, and named it New South Wales. After losing its North American colonies in 1783, Great Britain began deporting convicted criminals to the region in 1786. For the zoological and botanical efforts conducted on Cook's voyages see Rice 2000.

80. See Darwin 1838–1841, III, ix. Sulloway has cast lasting doubt on the accuracy of Darwin's place indications in *Zoology*. For the errors Darwin committed, see Sulloway 1982c, 61f. In terms of the observation that the islands of the archipelago were likely populated by different bird species, Darwin himself acknowledges in his text: "Unfortunately

I did not suspect this fact until it was too late to distinguish the specimens from the different islands of the group; but from the collection made for Captain FitzRoy, I have been able in some small measure to rectify this omission." See Darwin 1838–1841, III, 147.

81. See Darwin 1838–1841, III, ix.

82. See Darwin 1839, 475.

83. See Chambers 1844, 98. For the book's history and significance see Secord 2000; for the sales figures see 526 in particular.

84. Darwin obtained his own copy of *Vestiges*, a sixth edition, in 1847. However, as of 1844 he was aware of the book starting in 1844 from the extensive reviews in scientific journals. Most reviewers stressed the saltationistic tendency of Chambers's theory of evolution even more strongly than Chambers had in the book itself. Secord therefore writes: "Reading the reviews allowed Darwin to sharpen the distinction between his own theory and its competitors. *Vestiges* was reduced to a simplified version in which new organic forms sprang from old ones with no intermediate stages." Secord 2000, 431. For Darwin's reception of Chambers see Secord 2000, 426–436; Browne 1995, 445–447; and Desmond and Moore 1991, 316–320. Darwin himself also initially speculated about a saltationistic process in his notebooks; see Herbert 1980, 63 and 127. Also see Sulloway 1982b, 371.

85. See Di Gregorio and Gill 1990, 164, 148–149; and Secord 2000, 433.

86. The first publisher, Colburn, had neglected to pay Darwin his share of the books sold to that point. For Darwin's reproaches to Colburn see *Correspondence* 3, 175f. For more on the changes related to Tierra del Fuego see Darwin's letter to Murray in *Correspondence* 3, 204. In his *Narrative of the Surveying Voyages of H.M.S. Ships* Adventure *and* Beagle *Between 1826 and 1836*, which appeared at the same time as Darwin's account of the voyage, FitzRoy reported extensively on the Fuegians. For this reason Darwin initially limited his own account of them. In the second edition he expanded his observations but still did not include any pictures. Most of the new illustrations came from the geological works on the *Beagle* voyage that had recently appeared. The publishers Smith & Elder provided the plates free of charge. See *Correspondence* 3, 205 and 209. A lizard with a striking crest on its back came from Charles Lyell's *Elements of Geology* of 1838. See Darwin 1845, 385, and *Correspondence* 3, 229 and 243. For the skimmer see Darwin 1845, 137. In *Zoology* he describes the *Rhynchops nigra* but does not include an illustration. Darwin 1986 [1841], 143.

87. Regarding the illustration of *Rhynchops nigra*, Darwin informed Murray: "I thought it very desirable to illustrate a description of a curious

Bird: I will direct Mr. Lee to take it & the account to you, if you approve, if not I will pay myself." See *Correspondence* 3, 196. An entry in Darwin's account book from August 5, 1845, records the commission. See the editorial comment at *Correspondence* 3, 230. For John Lee see the same note. Speaking of the Galápagos finches, Darwin wrote to John Murray: "I have indulged myself in another woodcut at my own expence." See *Correspondence* 3, 230.

88. A longer passage on the finches can be found in the manuscript *Natural Selection*, which Darwin worked on at the end of the 1850s. Here Darwin writes: "Hence I suppose that nearly all the birds had to be modified, I may say improved by selection in order to fill as perfectly as possible their new places; some as Geospiza, probably the earliest colonists, having undergone far more change than other species; Geospiza now presenting a marvellous range of difference in their beaks, from that of a gross-beak to a wren; one sub-species of Geospiza mocking a starling, another a parrot in the form of their beaks." See Stauffer 1975, 257. Darwin, under pressure after Wallace sent him a letter in 1858 containing a draft of his theory of evolution that closely resembled his own, never finished the manuscript. Within a year he wrote a short version which appeared as *Origin of Species*. This work does not include the passage on the finches. See Sulloway 1982a, 35f.

89. Quoted in Darwin 1859, 21. For the position and history of the breeds of pigeons in Darwin's *Origin of Species* see Secord 1981 and Secord 1985. In the 1868 work *Variation Under Domestication* Darwin finally supplemented the extensive description from his earlier book with the pictures from his archive that he had previously failed to share with his readers. A letter to Huxley proves that Darwin was already in possession of the pigeon pictures at the end of the 1850s. See *Correspondence* 7, 428.

90. Darwin's parentheses. See *Correspondence* 11, 405. Salvin made his request to Darwin via Hooker; see *Correspondence* 11, 391. For the sales figures for the travel book see *Correspondence* 3, 381.

91. See Salvin 1876, 462. In this report, Salvin analyzes collections of bird specimens brought back by a traveler named Dr. Habel. The Galápagos islands had been visited before by other ships, but no significant bird collections had been obtained there. Shortly after Darwin's stop, the French research ship *Vénus* visited the Galápagos between 1836 and 1839, and the ship's doctor compiled a small collection of birds. The Swedish ship *Eugenie* also brought some of the islands' birds to Europe in 1852. Salvin writes: "The ground is classic ground; and the natural products of the Galapagos Islands will ever be appealed to by those occupied in investigating the complicated prob-

lems involved in the doctrine of the derivative origin of species." See Salvin 1876, 462.

92. See Salvin 1876, 484. For the tables with locations and measurements see Salvin 1876, 463–465 and 480–484.

93. For Ridgway's illustrations see Ridgway 1889, 106f; for Lack see Lack 1947, 57, and Table VIII. For Lack's role in popularizing the Galápagos finches see Sulloway 1982a.

94. For the numerous transformations from find to fact see Latour 1997.

95. For "adaptive variation" see Campbell and Lack 1985, 130ff; in terms of the finches in particular see Steinheimer and Sudhaus 2006.

96. Darwin advised Salvin not to be discouraged from this first impression: "Perhaps you may have read my Journal & if so, you will have observed that all the productions are singularly unattractive in appearance." See *Correspondence* 11, 404f.

Chapter 2. Darwin's Diagrams

1. See Darwin 1958, 117.

2. The point in time at which Darwin commissioned the printer to transfer the picture of the Galápagos finches to a woodblock is known, but not that at which he drew the picture itself. As a result, the sketch in Notebook B is considered the first picture of evolution.

3. See Notebook B, 36. Darwin's emphasis, in Barrett et al. 1987.

4. According to De Beer, Darwin read Malthus's essay in September 1838. See Oldroyd 1984, 334.

5. See Mivart 1873, 506.

6. François Jacob describes the role of "tinkering" or the playful combination of old and new materials in the history of science. For more on this principle, see Rheinberger 2003a. Heesen 2002 tracks this principle using newspaper reports. Dov Ospovat's trailblazing study of the textual level of Darwin's work clearly shows how evolutionary theory was generated from the topoi of natural history that preceded it. See Ospovat 1981.

7. For the dating of the Red Notebook, see Herbert 1980 and Sulloway 1982b, 371f. On Darwin's note-taking methods, see Herbert 1980, 5f. For the number of pages constituting Darwin's output, see Stott 2003, 65.

8. In 1842 Darwin acknowledged in a letter to his wife that he was working on a theory of evolution; in 1844 he first communicated this theory to a friend, the botanist Joseph Hooker. On Darwin's letters to his wife and Hooker, see Browne 1995, 452 and 446. On Darwin's close

contacts to the scientific world in London, see Rudwick 1982 and, in detail, Browne 1995.

9. On Owen's life and scientific work, see Rupke 1994.

10. Erasmus Darwin's *Zoonomia, Or the Laws of Organic Life* appeared from 1794 to 1796. It was primarily a medical treatise describing almost five hundred illnesses and their treatments. Although Erasmus Darwin was highly respected by his contemporaries, his theory of evolution had almost no influence on natural history in England. See King-Hele 2004, 21f.

11. On Jameson and Grant, see Secord 1991a. On Grant see Sloan 1985; Browne 1995, 80ff.; and Stott 2003, especially 1–41.

12. The article "Observations on the Nature and Importance of Geology" appeared in 1826. On its authorship and significance, see Secord 1991a.

13. See Lyell 1991 [1832], II, 1–175.

14. Darwin first read Lamarck during the year he studied with Grant in Edinburgh. See Sloan 1985, 76f. However, he found little inspiration from Lamarck for his theory of evolution, drawing instead on the work of his direct contemporaries. See Ospovat 1981.

15. See Darwin 1986 [1841], 63f.

16. See Darwin 1845, 379f.

17. See *Correspondence* 1, 513f. Desmond 1985 extensively describes the venomous disagreements among London's zoological societies and their political background.

18. On the surpassing of the museum in Paris, see Gunther 1975a, 106. The fact that the majority of specimens came to the British Museum was primarily thanks to the ambition of John Edward Gray, curator of the zoological department. The competing institutions were the Hunterian Collection of the Royal College of Surgeons and the Zoological Society. See Gunther 1975a, 62. On the history of London's zoological collections, see also Allen 1978 and Yanni 1999.

19. Both citations are reported in Tree 1991, 13f. Swainson lived from 1807 to 1814 in Sicily. Later, in the course of discussions with the natural history museum in Liverpool, he wrote a handbook on compiling and managing zoological collections. See Swainson 1822.

20. The first edition of *Systema Naturae* appeared in 1735, the twelfth and last in 1768. On the explosive growth in the collections, see MacLeay 1819–21, vi, and McOuat 2001, 481. Gunther 1975 empathically describes its impact on day-to-day museum work.

21. In 1812 Cuvier wrote: "The smallest bone surface, the tiniest Apophyse have a specific character in terms of the class, the order, the genus, and the species to which they belong, and to such an extent that

with the necessary skill and aided by a measure of clever application of analogy and actual comparison it is as possible to determine all the remaining parts from a single well-preserved end piece of a bone as if one had the animal itself before one's eyes." Translated from Jahn 1998, 327. On the characterization of organisms in Cuvier's theory as functional wholes that pass perfectly into their environments, see Ospovat 1981, 7ff. and 33ff. On Cuvier's idea of the concept of a species, see Lenoir 1989, 61–65.

22. On the mechanistic teleological view of nature in Germany, see Lenoir 1989 and the chapter "The Mechanism of Formation" in Larson 1994, 132–169. Lenoir considers Carl Friedrich von Kielmeyer to be a pioneer of the embryological method. See Lenoir 1989, 41.

23. On Herschel, see the chapter "Of the Classification of Natural Objects and Phenomena, and of Nomenclature" in Herschel 1831, 135–143. On Whewell see the chapter "Classificatory Sciences" in Whewell 1837, 286–414.

24. On these astronomy metaphors see Winsor 1976, 175f. On the state of English physics see Babbage 1830.

25. McOuat's outstanding study traces the discussion about the assigning of taxonomic names during the first half of the nineteenth century. See McOuat 1996. O'Hara 1988 and O'Hara 1991 examine the visual aspect of classification systems in the nineteenth century, as do Di Gregorio 1982 and Di Gregorio 1987.

26. See DAR 205.5.40; also see Ospovat 1981, 101.

27. The word *paragone*, literally "comparison," is used to indicate a competition among the arts. In addition to disputes about the primacy of painting or sculpture, Lessing's "Laocoon" essay brought the comparison between visual arts and poetry into focus in the eighteenth century. On the clash between image and text in the sciences, see Koebner 1989, Mazzolini 1993, Baxmann 2000, Latour and Weibel 2002, and Daston and Galison 2002. On the heated debates about the natural system, see McOuat 1996 and McOuat 2001. Also see Winsor 1976, Di Gregorio 1982, and Desmond 1985.

28. Quoted in McOuat 2001, 11.

29. See McOuat 1996, 191 and 199.

30. See McOuat 1996, 499.

31. On Strickland and the nomenclature reform see Jardine 1858, Di Gregorio 1987, and McOuat 1996. Its significance for Darwin's understanding of the concept of "species" is most extensively discussed in McOuat 1996. *The Dodo and Its Kindred* appeared in 1848.

32. Here Strickland adopts a position that Linnaeus had already formulated but less consistently applied. Linnaeus compared names

of species with coins, saying that both items acquire their value solely from their circulation. See Müller-Wille 1999 and Müller-Wille 2002, 53.

33. See Strickland 1835, 40. On the further history of the rules of nomenclature see Heppel 1981. Biologists continue to disagree today about how species should be named. See Mayr [1942] 1982 and Mayr 1996.

34. See Strickland 1841, 185.

35. See Strickland 1841, 187, 192, and 190.

36. On classificatory pictures of the order of birds in the nineteenth century, see O'Hara 1988. Strickland provides a report on the annual meeting in Glasgow in a letter. See Jardine 1858, cciii.

37. See Swainson 1837, ii and 199. The author is quoting from an essay by M. M. Horsefield and A. Vigors in *Linnean Transactions* XV, 328. On his thirteen years of research, see Swainson 1836, II, 198. On Swainson's criticism of Cuvier's binary system see Swainson 1836, II, 193.

38. Ospovat 1981 extensively examines the connection between Mac-Leay and Darwin. On MacLeay's influence on English taxonomy in the nineteenth century, see Winsor 1976. Agassiz provided a brief description of MacLeay's system, see Agassiz 1962 [1859], 241–245. The Zoological Society was a quinarian stronghold, see Desmond 1985, 161f. On the significance for ornithology see Stresemann 1951, 176f. As a representative of the British crown, MacLeay lived primarily abroad as of 1825 and died in 1865 in Australia.

39. On "the butterflies of the vertebrates" see Swainson 1836, i and 6. For the citation "No rational being can suppose that the great Architect of the world has created its inhabitants without a plan," see Swainson 1835, 319. The remark on the animal kingdom and planetary orbits is reported in Stresemann 1951, 177.

40. See MacLeay 1818–1821, 395.

41. A postscript of this kind appears, for example, in Vigors 1838, 509. However, the history of strained relationships between authors and lithographers, engravers, etc., is much older. Among naturalists, Linnaeus clearly testified to this situation when he said that every illustrator of plants should be supervised by a botanist. See Müller-Wille 2002, 8.

42. On Wallace, see Wallace 1856, 207. The quinarians were known for producing floods of diagrams at their meetings, of which only a very few were ever printed. See Desmond 1985, 162.

43. On the dismissive remark on the naming of species, see MacLeay 1819–1821, xvi–xvii f. Strickland draws on a motif first formulated by

Linnaeus to counter his contemporaries' preference for symmetry: discussing kinship among plants, Linnaeus said that the relationships among them resemble a territory on a map. See Rheinberger 1986, 240; also see "Das Natürliche System als Karte," in Müller-Wille 1999, 89–104. At the close of the eighteenth century a series of diagrams were produced that acknowledge the asymmetry in nature, but these apparently received little attention in England. On this and on the maplike form of the natural system in general, see Barsanti 1992.

44. See Strickland 1841, 187 and 190. On the history of the tree as an organizational metaphor in zoology see Barsanti 1992.

45. Darwin's emphasis; Notebook B, 21.

46. See Darwin 1859, 48. Also see Sulloway 1982b, 379. The sense that the distinction between species and variety is often random provides a connecting thread throughout *Origin of Species*. On the species of pigeons, Darwin wrote: "Altogether at least a score of pigeons might be chosen, which if shown to an ornithologist, and he were told that they were wild birds, would certainly, I think, be ranked by him as well-defined species. Moreover, I do not believe that any ornithologist would place the English carrier, the short-faced tumbler, the runt, the barb, pouter, and fantail in the same genus; more especially as in each of these breeds several truly-inherited sub-breeds, or species as he might have called them, could be shown him." See Darwin 1859, 22f.

47. See Steinheimer and Sudhaus 2006.

48. On the correspondence of Darwin and Strickland in 1842 and 1849 see *Correspondence* 2 and *Correspondence* 4. On Darwin's early will see Browne 1995, 447.

49. On Darwin and MacLeay see Ospovat 1981, 101–114.

50. On William Smellie and MacLeay see Stresemann 1951, 176. On Darwin's notations, see Chapter 1.

51. See Notebook B, 23, in Barrett et al. 1987. On the motif of the tree in natural history, see Lam 1936, 156, and Barsanti 1992.

52. See Notebook B, 20 and 26.

53. On Darwin's use of visual elements drawn from geology see Di Gregorio 1982, 247. On Darwin's close involvement with geology see Secord 1991a, Herbert 1991, and Herbert 2005.

54. See Notebook B, 26.

55. Oppenheimer and, most recently, Richards claim there can be no doubt that Darwin read Barry's article shortly after its publication. Richards also cites the similarity to Barry's diagram as evidence. See Richards 1992, 108f.; also see Oppenheimer 1967, 245. In his copy of Richard Owen's *Lectures on the Comparative Anatomy and Physiology of the Invertebrate Animals*, which he owned in the second edition of

1855, Darwin summarized Owen's discussions with the note "Barry" in the margin. See Di Gregorio and Gill 1990, 653. Martin Barry's familiarity with German embryology was based on personal contacts. In 1833, he had studied physiology and anatomy in Heidelberg under Friedrich Tiedemann, a student of Blumenbach's. In 1839 he was awarded the Royal Medal for his service in the field of embryology. See Oppenheimer 1967, 245. On Barry's influence on embryology in England, see Oppenheimer 1967, 242–245; Ospovat 1981, 11 and 124f.; Richards 1987, 134; and Richards 1992, 108–111.

56. Barry's emphasis; see Barry 1836–1837, 121. For more on classification, see Barry 1836–1837, 136f.

57. See Chambers 1844, 213 and 222. On the history of the idea that ontogeny mirrors phylogeny, see Richards 1976, Richards 1992, and Hopwood 2006.

58. See Di Gregorio and Gill 1990, 164.

59. See Barry 1836–1837, 116.

60. See also Desmond 1982. Commenting on Richard Owen's *On the Nature of Limbs* from 1849, which was strongly influenced by German embryology, Darwin wrote on the book jacket: "I look at Owen's Archetype as more than ideas, as a real representation." See Di Gregorio 1990, 655, and Chapter 3.

61. On the comparison among Barry, Carpenter, and Chambers see the excellent analysis in Richards 1987.

62. On Haeckel see Haeckel 1866, 32. For Darwin's citation see Darwin 1859, 433.

63. On the central role of pictures in geology see Rudwick 1975, Rudwick 1992, and Rudwick 2000.

64. See Darwin 1859, II, 140. On Darwin's cross-sectional diagrams, see Darwin 1846.

65. When Darwin suggested in 1859 that three hundred million years had passed since the disappearance of the dinosaurs, this assertion was quite daring. Since that time assumptions about the age of the earth have skyrocketed upward. See Darwin 1859, 287. On the debate about the date proposed by Darwin, see Burchfield 1974.

66. Louis Agassiz was born in Motiers, Switzerland, and studied science in Zurich, Heidelberg, and Munich, before traveling to Paris in 1832 to study under Cuvier. Cuvier died a few months later. In his will Cuvier named Agassiz as heir to some of his scientific work, and the younger scientist thus considered himself to be Cuvier's successor. Agassiz also enjoyed the intensive support of Alexander von Humboldt, who arranged a professorship for him at the Collège de Neuchâtel. In

1846 Agassiz went to Harvard, where he remained until his death. On Agassiz's life and work see Agassiz 1962 [1859] and Winsor 1976.

67. Agassiz was the first to use this method of presentation in the field of paleontology. See Gould, Gelinski, and German 1987, footnote 7, 1441.

68. Agassiz maintained a particularly strict creationist position that assumed multiple divine interventions during the history of the natural world. He went so far as to claim that identical species that occurred in different regions came about in separate acts of creation. See Agassiz 1962 [1859], 252–302, and Agassiz 1874. His division of the animals into *embranchements* came from Cuvier, who used four organizational principles to distinguish them: the vertebrates have a spine; the mollusks have a brain, but no nervous system; the insects have a small brain and two different nerve strands, and the last group can be identified by the radial symmetry of their structure. On Agassiz's classification system, see Lurie's introduction to Agassiz from 1962 [1859], iv–xxxiii, Winsor 1991, 19–22, and Ruse 2003, 62ff.

69. See Darwin 1872, 310. Darwin had earlier evoked Agassiz's method of depiction in the fossil maps by suggesting the possibility of representing "the number of the species of a genus, or the number of the genera of a family" with "a vertical line of varying thickness, crossing the successive geological formations in which the species are found." See Darwin 1859, 316f. Martin Kemp has also pointed out the similarities between Agassiz's drawing and Ernst Haeckel's genealogical tables from 1868's *Natural History of Creation*. See Kemp 2000, 91. See also Brink-Roby 2009.

70. Darwin's sketches with the draft diagrams can be found in DAR 205.5–6. On Darwin's experiments, see Rheinberger and McLaughlin 1984. On the findings of his statistical analyses on the subject of divergence among species see Browne 1980. On the barnacles see Stott 2003.

71. See *Correspondence* 7, 107. On Wallace's letter and its consequences see Shermer 2002, 88 f.; also see Raby 2001, 137f. and Browne 2002, 14–42. The claim that Darwin copied parts of Wallace's theory or lied about the date he had received the letter has found no defenders among recent scientific historians. Brackman 1980 proposes this thesis. Wallace's original letter has been lost.

72. See Wallace 1855.

73. See Wallace 1856, 205. For the instructions for preparing diagrams, see Wallace 1856, 206f. Concerned that lithographers could inadvertently alter diagrams during the artistic process, Wallace advised naturalists to use only simple symbols commonly available to printers.

74. See Wallace 1856, 206.

75. See Wallace 1858, 36.

76. Lyell had already made Darwin aware of Wallace in April 1856. See Shermer 2002, 89. In December 1855 Edward Blyth likewise mentioned Wallace's essay to Darwin. See *Correspondence* 5, 519f. The marginal notices to Wallace 1855 follow the citations in Shermer 2002, 89.

77. On the tree comparison see Wallace 1855, 8. On Wallace's reading of Malthus see Jones 2002, 86f.

78. Quoted following Browne 2002, 42.

79. The long manuscript of the *Big Species Book* was published in Stauffer 1975. The explanations of the diagrams appear in Stauffer 1975, 236–246.

80. On Darwin's comment about the reader's "convenience of reference," see Stauffer 1975, 238. For the instruction to the engraver see DAR 10.2.26s.

81. See Darwin 1872b, 90.

82. See Darwin 1872b, 59.

83. On the lack of a point of origin, see Müller-Wille 2005.

84. See Darwin 1872b, 91.

85. This time period is specified in the first English edition only. See Darwin 1859, 287. On the heated debate that this comment inspired, see Burchfield 1974.

86. On Haeckel see Di Gregorio 2005, Kleeberg 2005, and Richards 2008.

87. On Strickland's shrubs, see Strickland 1841, 190, and the discussion in this chapter. On Agassiz's map, which also appears in Haeckel's later drawings, see Kemp 2000, 91.

88. Stephen Jay Gould also criticizes this illustration because Haeckel, following his preferences, devoted too much space to the carnivores and primates. However, it seems even more interesting that Haeckel places humans at the same level as the other mammals in this drawing. See Gould 1989, 265. Furthermore, Richards points out that Haeckel draws a distinction between "the paleontological developmental history of the direct progenitors of each individual organism" and the "true-stem-tree," i.e., of the type that appears in his *Generelle Morphologie*. In *Anthropogenie*, in contrast, he examines man's immediate ancestors and therefore depicts a lineal descent. See Richards 2008, 138.

89. Quoted in Browne 2002, 270. He failed to mention that he himself had already drawn other ancestral trees, such as that of the mammoths. See *Correspondence* 8, 379f.

90. Haeckel's family trees were the objects of much criticism. See Gould 1996, 67. However, Gould mistakenly claims here that Haeckel's oak appeared in *Generelle Morphologie*. On criticism of Haeckel, see also Bowler 1996, 425f., and Bowler 1992, 85f.

91. As Peter Bowler aptly explained: "The tree of evolution was thus redrawn to hide the ladder within, providing an axis for the progress that was widely hailed as a necessary feature of the history of life." See Bowler 1996, 425. On the image of the ladder in the history of biology, see Barsanti 1992 and Lovejoy 1964.

92. See Mivart 1871, 97 and 113. Mivart admits that he took the comparison from Galton 1869.

93. See Owen 1860, 407.

94. The expression has never been found in Herschel's writings, but the remark was reported to Darwin in 1859. See *Correspondence* 7, 423.

95. For the German tradition of incorporating a teleological perspective into natural history, from Kant to Haeckel, see Lenoir 1989. On English teleology see Ospovat 1981.

96. See Darwin 1872, 291. For criticism of this idea see Karl Ernst von Baer's 1873 book *Über Darwins Lehre* and the extensive commentary on it in Lenoir 1989, 254. The law of "natural selection" found little acceptance even among the small circle of Darwin's closest fellow scientists, such as Huxley or Haeckel, who similarly rejected indeterminate becoming for directed development. See Bowler 1992, 76f.

97. Between 1859 and 1861 William Charles Linnaeus Martin, whom Darwin had earlier consulted on questions of animal husbandry, sent a draft of a genealogy of the birds. For the text and image, see the letters from William Charles Linnaeus Martin to Charles Darwin, 1859–1861, in *Correspondence* 13, 399–410.

98. See *Correspondence* 13, 402.

Chapter 3. The Picture Series

1. See DAR 84.150.

2. See DAR 84.

3. On the cooperation between Darwin and Ford, see Gunther 1972 and Gunther 1975b.

4. In the first edition of *Descent of Man* the series consisted of just four pictures; in the second, five (Figure 60 had been added). However, Darwin had apparently planned to add an illustration when preparing the first edition, as he writes, "I regret that I have not given an additional drawing, besides fig. 58, which stands about halfway in the series

between one of the simple spots and a perfect ocellus." See Darwin 1871, 147. All further details here come from the second edition.

5. See Darwin 1882, 434.

6. See Darwin 1882, 397, Figure 51; 399, Figure 52; 431–440, Figures 54–61.

7. See Sternberger 1977, 106.

8. Criticism is often leveled at the simplification inherent in newer images of evolution, which, often enough, present the process as a history of progress. See Gould 1996 and Bowler 1996, 419–446.

9. As examples from the many introductions to biology or evolutionary theory, see Strickberger 2000; Storch, Welsch, and Wink 2001, 343; Mayr 2003, 36, 48, 75, 252, 296.

10. For more recent attempts to explain striking ornaments found among animals in terms of evolution, see Grammer and Voland 2003 and Gadagkar 2003. Menninghaus 2003 is one example of an attempt to analyze and historically trace the various positions in this debate. The question also featured prominently in the Endless Forms exhibition. See Munro 2009.

11. The metaphor of the net is used repeatedly in the history of science to describe the connections in the production of knowledge. See Star and Griesemer 1989 and Latour 1990.

12. See Darwin 1882, 421.

13. See Cuvier 1805, Plate XI, "Muscles et os de la main."

14. Owen introduced the word *dinosaur* at an 1841 meeting of the British Association for the Advancement of Science in Plymouth. See Rudwick 1992, 141f. On Owen's concept of the archetype, see Richards 1987 and Rupke 1995. The table appears in Owen 1848 and Owen 1849a.

15. See Owen 1849a, 86.

16. On the passages Darwin underlined, see Di Gregorio and Gill 1990, 484–485. The French edition was published from 1806 to 1809 by the physician Jacques-Louis Moreau de la Sarthe. See Arburg 1999, 40–59, and Schlögl 1999, 164–171. On physiognomy in the nineteenth century see Schmölders 1995, Campe and Schneider 1996, Hartley 2001, and Schmidt 2003. On the potential link between evolutionary theory and physiognomy see Brednow 1969 and Clausberg 1997.

17. Lavater's contemporary Petrus Camper, a Dutch anatomist and physician, arranged frontal views of human skulls into visual tables in a way that is often read as a sequence in time representing a racist history of evolution. Meijer, however, contests the idea that this was Camper's intention. On this issue and the reception of Camper in the field of evolutionary theory, see Meijer 1999 and Hagner 1999.

18. On the table, see Ecker 1851–1859. Ecker reports using illustrations from other works for those stages for which no specimen was available, such as those of the French zoologist Coste or pictures of wax models from the Ziegler company. On the introduction of picture series to developmental biology and the standardization of the linear sequence in the eighteenth century, see Ratcliff 1999. On the visual history of embryology see Wellmann 2003 and Hopwood 2000, 2002a, 200b, 2005, and 2006. On the history of embryology see Duden 2002.

19. See Darwin 1872, 310.

20. See Darwin 1871, I, 15f. Also see Hopwood 2006, 287f.

21. See Darwin 1871, I, 15, Figure 1. The human embryo was Figure 2 from Ecker 1851–1858, Plate XXX. The dog embryo comes from Bischoff 1845, Plate XI, Figure 42B. The physiologist Alexander Ecker published his work *Icones Physiologicae* between 1851 and 1859. Previously, Theodor Ludwig Wilhelm von Bischoff had published *Entwicklungsgeschichte des Kanincheneies* (Developmental History of the Rabbit Egg) (1842) and *Entwicklungsgeschichte des Hunde-Eies* (Developmental History of the Dog Egg) (1845).

22. See Di Gregorio and Gill 1990, 655. On Darwin's interpretation of the archetype as a real historical phenomenon, see Desmond 1982.

23. See *Correspondence* 11, 148. Argyll, quoted in Desmond 1994, Figure 29, between pages 350 and 351. On the debate with Bishop Wilberforce see Desmond 1994, 276–281. On the book's public reception see Desmond 1994, 312–335, and Lyons 1999, 189–226.

24. On Huxley's illustrations see Bodmer 1991. The artist and sculptor Benjamin Waterhouse Hawkins had previously drawn the fish and reptiles for Darwin's *Zoology of the Voyage of the H.M.S. Beagle*. His most famous commission was to create the dinosaur sculptures in the Crystal Palace Park in 1852. On Hawkins see Mitchell 1998.

25. An article in the weekly journal *Athenaeum* reports on "a photograph made by Mr. Fenton for the Trustees of the British Museum (from the skeleton of the animal [i.e., the gorilla] in the British Museum), which is now sold, at the South Kensington Museum, for a few pence." Gray 1861b, 695. On this history of the photograph see Lechtreck 2008.

26. See Huxley 1863, 97.

27. In the twelve editions of *Systema Naturae* published between 1735 and 1768, Linnaeus revised the classification of the apes several times. In 1758 he replaced the family designation *anthropomorpha* with *primates:* this category included *Homo, Simia, Lemurae* and *Verspertillo* (bats). Among the criteria for distinguishing apes and human in his classification scheme is human reason (ratio). See Spencer 1995,

16. On Linnaeus's difficulty in ultimately distinguishing between humans and apes, see Agamben 2004, 24f.

28. On Huxley's illustration, see Haeckel 1874, 488. It appears again in a modified form in Haeckel's *Das Menschenproblem und die Herrentiere von Linné* (The Problem of Man and Linneaus's Primates) of 1907. See Gursch 1981, 110. On the picture of the human developmental series see Gould 1996 and Mitchell 1998.

29. On Philipon's famous 1831 pear caricature see Clausberg 1997 and Bosch-Abele 2000. On Grandvilles *Métamorphoses* from the 1830s and 1840s see Grandville 2000 and Kerr 2000. On human/animal metamorphoses in art and science see Clair 1993.

30. On caricatures of Darwin see Baumunk and Rieß 1994 and Browne 2001. Kingsley's *The Water-Babies* appeared in 1863 in a first edition containing three illustrations; the second edition, from 1885, included a hundred illustrations by Sambourne. See Kingsley 1984 [1885].

31. Quoted in Browne 2001, 507. Darwin's folders of caricatures can be found under DAR 140 and DAR 225, 175–185.

32. See Mandelstam 1991 and the introduction to this book.

33. See Di Gregorio and Gill 1990, 447 (on Lamarck) and 164 (on Chambers).

34. See Darwin 1872b, 265.

35. Haeckel presented the first comparative plates with two embryos each in 1868's *Natural History of Creation.* The expanded plates showing three stages appeared in 1874 in *Anthropogenie oder Entwickelungsgeschichte des Menschen;* see Hopwood 2006. On the idea of temporalization see Lepenies 1976. Darwin refers to the archaeopteryx only in passing; see Darwin 1882, 181, and Darwin 1872b, 312.

36. See Darwin 1882, II, 434.

37. See Darwin 1903, I, 330. Darwin corresponded with Thomas Woolner, the sole Pre-Raphaelite sculptor, and sat for him for a portrait in November 1868. On the Pre-Raphaelites' image of nature see Wullen 2004.

38. See Paley 2005 [1802], 19f.

39. Paley's book appeared in seven editions between 1802 and 1804. It was translated into French in 1804 and into German in 1837. Ruse 2003 provides an overview of the history of the argument from design from the seventeenth century to the present. On the perception of nature as art before Darwin, see Bredekamp 1995. On the topos of nature as an artist see Daston and Park 2002, 341–354.

40. See Chapter 1. The most thorough consideration of the question of how the tradition of natural theology influenced Darwin throughout his life can still be found in Ospovat 1981. On the title of the series see Ruse 2003, 44.

41. See Whewell 2000 [1846]. In Whewell's opinion, the natural phenomenon that best demonstrated the work of a Creator was the kangaroo's nipple. See Whewell 2000 [1845], 123ff. The short tract *Indications of the Creator* contains a condensed version of excerpts from the author's previously published works. See Lynch's introduction to this volume.

42. See Ruse 2003, 80, and Secord 2000, 227–229.

43. A number of available popular works trained a lens of natural theology onto local flora and fauna. See Allen 1994 and Lightman 2000.

44. See Dickens 1851, 291. Gillespie 1990 considers the link between Paley's view of nature and industrial design.

45. On the exhibition see Tree 1991, 174, and Dickens 1851, 289.

46. On the use of feathers in Victorian fashion see Briggs 1988. The use of miniature worlds to represent colonial power is described by the historian Timothy Mitchell based on the Egyptian city built for the 1889 World Exhibition in Paris. See Mitchell 1989. On the relation of biogeography and empire see Browne 1996.

47. On the exhibition see Blunt 1976, 58–60. On Gould's ornithological collection see Tree 1991, 157 and 173. On Darwin in London see Stott 2003, 183f.

48. All quotes are from Gould's introduction to the hummingbirds; see Gould 1849–1861, xxviii. On Buffon's enthusiasm for the hummingbirds see Spary 1999, 112.

49. Quoted in Tree 1991, 164. The patent was subsequently disputed when the American William Lloyd Baily accused Gould of stealing his process, which he had described in a letter. See Tree 1991, 165; also see Gould 1849–1861, i, vii.

50. William Swainson's *Zoological Illustrations* appeared between 1820 and 1832. Seventy of the one hundred and eighty-two plates show birds. See Tree 1991, 37.

51. In the introduction Gould refers to their contributions in his typically vague manner: "I am especially indebted to those persons connected with the production . . . to my artists Mr. Richter, to Mr. Prince, and to Mr. Bayfield" See Gould 1849–1861, I, vii.

52. See Gould 1849–1861, I, xviii.

53. On the list of subscribers see Gould 1849–1861, i.

54. See Tree 1991, 164.

55. See Tree 1991, 158.

56. The complete citation is: "It is to these men, living and dead, that science is indebted for a knowledge of so many of these 'gems of creation.'" See Gould 1861, ii. On Darwin's markings in the work, see Di Gregorio and Gill 1990, 342–344.

57. See Gould 1849–1861, i, xii.

58. Gould wrote: "The gorgeous colouring of the Humming Birds have been given for the mere purpose of ornament, and for no other purpose of special adaptation." See Gould 1849–1861, I, xii, xvii. Newton is quoted in Tree 1991, 59.

59. On Darwin's note, see Di Gregorio and Gill 1990, 342. On the letter, see *Correspondence* 9, 295.

60. In *Descent of Man*, Darwin briefly comments on Gould in the section "Variability of Birds." See Darwin 1882, 413. Some manuscripts also seem to suggest that Darwin conducted color studies. A number of theories to explain why birds were brightly colored circulated during the 1860s; see Gould 1849–1861, i, xxxvii. On Darwin's optical experiments see DAR 84.2.28–30.

61. Argyll had published in the *Edinburgh Review, Good Words,* and the notices to the Royal Society of Edinburgh. See Argyll 1867, foreword. Wallace's article "Creation by Law" was published in 1867 in *The Quarterly Journal of Science.*

62. With *The Reign of Law* (1867) and *Primeval Man* (1869), Argyll published two books critical of evolutionary theory within a short time. In 1884 *The Reign of Law* was already in its fifth printing, and in that same year Argyll published a second book with the same argument, *The Unity of Nature.* His role in the debate surrounding Darwin has been the subject of little research. See Gillespie 1977. Currently the most informational sources for following the duke's career are Ellegard 1990 and Desmond and Moore 1991.

63. See Charles Darwin to Charles Kingsley on June 10, 1867, in *Correspondence* 15, 297. On Wallace see Wallace 1867, 471.

64. Argyll writes: "[Darwin's] theory gives an explanation, not of the process by which new Forms first appear, but only of the process by which, when they have appeared, they acquire a preference over others, and thus become established in the world." See Argyll 1867, 229.

65. See Argyll 1867, 247f.

66. See Argyll 1867, 202f.

67. See Wood 1867, 607.

68. On the history of trompe l'oeil in art see Gombrich 2002 and Ebert-Schifferer 2002. On pictures "painted" by nature on minerals, animal pelts, or butterfly wings, see Elkins 1999, 236–251. On the history of mimicry research see Kimler 1983.

69. See Argyll 1867, 221.

70. The bourgeois origins of the theory of evolution are attested to by the fact that ideological derivations of its teachings emerged in both conservative middle-class and socialist circles, but were never em-

braced by a monarchist group. On the broad spectrum of the theory's reception see Glick 1988 and Numbers 1999. On the group of English art aficionados that coalesced in the eighteenth century into a "gentleman's culture," see Dobai 1974. On the debates about beauty in the animal kingdom, see Menninghaus 2003.

71. See Argyll 1867, 104. For further examples, see Argyll 1867, 128f. (flight mechanism of birds), 195 (praying mantis), 202 and 243f. (hummingbirds), and 200 (lichens).

72. See *Correspondence* 7, 388. *Origin of Species* appeared on November 24, 1859. Ospovat 1981 and Browne 1980 have shown at length how strongly Darwin clung to natural theology's idea of ultimate purposes, which organisms were created to fulfill. For an encapsulated discussion, see Ruse 2003, 91–106.

73. In Darwin's words: "If it could be demonstrated that any complex organ existed, which could not possibly have been formed by numerous, successive, slight modifications, my theory would absolutely break down." See Darwin 1859, 189.

74. See *Correspondence* 8, 140. Roberts 1997 summarizes and analyzes the doubts that Darwin repeatedly expressed in his letter to Gray.

75. See Paley 1802, 49. Quoted in Ruse 2003, 43f.

76. A mass of notes, journal clippings, engravings, and test prints are stored with the drawings. Some of the notes contain the notation "British Museum." See DAR 84. The Darwin archive also includes a number of "sample packets," which are envelopes containing various types of feathers. See DAR 142.48–54. The letters from his correspondents can be found under DAR 89.81.

77. Ford was an artist in the zoological department of the British Museum. He was recommended to Darwin by department head Albert Gunther. See Gunther 1972 and Gunther 1975b.

78. See Darwin 1882, 434.

79. For Darwin's citation, see Darwin 1882, 420f. On the illustrations in *The Field* see Tegetmeier 1874. On the question of beauty see footnote 10.

80. On secondary characteristics see chapters XIII, XIV, XV, and XVI in Darwin 1882. The argus pheasant appears in chapters XIII and XV. Chapters VIII–XVIII are about sexual selection in animals.

81. Murray, who was concerned about the book's success, wrote the following in a letter to Darwin: "It is with a view to remove any impediments to its general perusal that I wd. call your attention to the passage respecting the proportion of advances made by the two Sexes in Animals. I wd. suggest that it might be toned down—as well as any other sentences liable to the imputation of indelicacy if there be any"; John

Murray to Charles Darwin, September 28, 1870, DAR 171. Quoted in Browne 2002, 349. Darwin coined the term "sexual selection" in 1856 while examining the differences between ethnic groups. See Desmond and Morris 2009, 282.

82. On hummingbirds see Darwin 1882, 414; on humans see Darwin 1882, 581. The concept of race in Darwin still requires study. It seems likely that Darwin did not define race in biological terms.

83. On birds, see Darwin 1882, 372, Figure 39 (woodhen *Tetrao cupido*); 373, Figure 40 (umbrellabird); and 382, Figure 46 (bower birds); on beards, see Darwin 1882, 381.

84. See Darwin 1882, 421. Surprisingly, Darwin does not raise the question of whether a bird's eye perceives things in an essentially different way than a human eye does.

85. On the citations from Darwin see Darwin 1882, 440f. Darwin added the reference to Helmholtz in the second edition. On the passage cited by Darwin, see Helmholtz 1871, 22f. On the reception of Helmholtz's studies of vision and the eye in Victorian literature, see Beer 1996. On the diverse reception of Helmholtz in the arts and sciences see Crary 1990 and 2002.

86. See Darwin 1882, 437f.

87. See Darwin 1882, 15, Figure 2. The shape of the ear also serves as a means to identify the creator and origin of a work of art in the Morelli method, developed in 1874 by Giovanni Morelli (alias Ivan Lermolieff). Ginzburg fails to mention this parallel, which warrants further investigation. See Ginzburg 1983.

88. Latour 1997 clearly lays out the steps necessary for moving from the object itself to its scientific representation.

89. See Darwin 1958, 47, and Darwin 1903, I, 398.

90. Kleeburg 2005 thoroughly examines the role of teleology in Haeckel's theory of evolution. On Haeckel's Zeus and Lavater's Apollo, see Clausberg 1997. On the feather as a favorite object of Darwin's, see Darwin 1903, 14f.

Chapter 4. The Laughing Monkey

1. See Savage and Wyman 1847.

2. See Du Chaillu 1861b, 60 and 71.

3. See Du Chaillu 1861b, 208.

4. Du Chaillu considered Africans to be an "inferior race" and predicted that they would die out for this reason. See Patterson 1974, 660f., and Browne 2002, 156f.

5. *On the Movements and Habits of Climbing Plants* appeared in 1875, *The Effects of Cross and Self Fertilisation in the Vegetable Kingdom* in 1876, *The Different Forms of Flowers on Plants of the Same Species* in 1877, and *The Formation of Vegetable Mould, Through the Action of Worms* in 1881. Along with these new publications, Darwin worked during the 1870s on new editions of *Origin of Species* and *Descent of Man.* For the quote, see Darwin 1959, II, 349.

6. See Ekman's introduction to Darwin 1998 [1872], xxiv.

7. See Darwin 1882, 79.

8. On the lady baring her teeth see Darwin 1872b, Figure 1, on Plate IV. On the laughing monkey see Darwin 1872b, 136, Figures 16–17. On the connection between *Expression of Emotions* and *Descent of Man* see Darwin 1958, 131.

9. See Darwin 1872b, 13f. Also see Browne 1985a, 153.

10. On Donders, Duchenne, and Bell see Darwin 1872b. Also see Di Gregorio and Gill 1990, 47f. (Bell), 204 (Donders), and 210ff. (Duchenne). On Donders's idea that crying is a protective mechanism see Darwin 1872b, 176.

11. See Notebook M, 146, 147.

12. See Notebook M, 85 (yawning), 129 (pouting orangutan), 137 (sulky pig-tailed baboon), 140e (Jenny ashamed of herself), and 142 (attempt at classification). For the record of his son William's development, see Darwin 1877. See also Browne 1985b.

13. See Notebook N, 6; on editions of Lavater in England see Graham 1979, 62.

14. See Darwin 1958, 72 and 79. On Lavater's popularity in the nineteenth century, see Schmölders 1995.

15. See Notebook N, 10. On Darwin's notes, see Di Gregorio and Gill 1990, 484f. The irregular underlining shows that Darwin skipped over entire volumes.

16. On Lavater's theory of natural signs see Schlögl 1999. On his attempt to establish physiognomy as a science, see Hagner 2004, 36–41, and Arburg 1999.

17. On cats see Lavater 1968 [1775–1738], II, 260. On elephants see Lavater 1968 [1775–1738], II, 154f. The theory that the form of every animal corresponds to its ruling passions, and that these same features recur in human faces, originated with the Italian Giovanni Baptista della Porta, who published the richly illustrated tract *De humana physiognomia* in 1591. Lavater criticizes della Porta at numerous points, primarily because, as he claims, the human face is differently proportioned and utmost caution is required when identifying similarities

between humans and animals. However, Lavater is just as convinced as della Porta of the lack of ambiguity in animal physiognomy. See Mraz and Schlögl 1999, 384.

18. On snarling, see Darwin 1872b, 250; on smiling, see Darwin 1872b, 134.

19. Michael Neve has pointed out that this lack of typologizing is also evident in *Expression of Emotions:* "There were no dividing lines between taxonomic categories, no differences between species and varieties, no in-built properties." Quoted in Browne 1985a, 161.

20. Bell's work appeared in several editions, the fourth and last in 1847. Darwin owned the 1844 third edition. No studies exist of the reception of Bell within artistic circles or the educational system. Bell wrote the book in the hope that it would earn him a position at the Royal Academy of Arts, but this plan did not come to fruition. See Browne 1985b, 163.

21. See Bell 1847, 126. On the criticism of antique art see Bell 1847, 10f., and on Rubens's lions see 123f.

22. On human physiognomy see Bell 1847, Plate II and 271. On the physiognomy of the dog see Plate III and 273.

23. On the limitations of animal faces see Bell 1847, 138. Darwin mentions this as a thesis to be refuted in Darwin 1872b, 11. On Bell's editorial role see Paley 1836.

24. On the laughing dogs see Notebook M, 84e. On the timing of his reading see Darwin 1958, 132. His reading notes list the period between June 10 and November 14, 1840. See Browne 1985b, 163.

25. See Bell 1847, 69. On Darwin's marginal note, see Di Gregorio and Gill 1990, 48.

26. See Correspondence 15, 141.

27. See Darwin 1872b, 23. The variety of the facial muscles and the resulting need for abstract depiction had already been emphasized by the Göttingen anatomist Albrecht von Haller in his two-volume *Icones Anatomicae* of 1756. See Daston 2004, 167.

28. See Darwin 1872b, 23.

29. Prodger 1998a, 4. The hunt for pictures in London's shops occurred around the same time that Darwin's notes record his visits to the British Museum. It follows that Darwin spent more time in the capital than his image as "the scholar from Downe" would suggest. See Browne 2003. On the trip to London see *Correspondence* 14, 358.

30. On the position of Darwin's research in his work overall, see Browne 1985a, Browne 1985b, and Burckhardt 1985. On Polly see Browne 2002, 361. Darwin's questionnaire is reprinted in 1872b, 15–16.

31. See Browne 2002, 298–303.

32. On Victorian photography see Bartram 1985 and Smith 1995. On Carroll's photo of Darwin see Prodger 1998a, 70.

33. See *Correspondence* 12, 240f. On the portrait's prophetlike stylization of Darwin, see Browne 2003 and Voss 2008.

34. Darwin's failure to include photos by Cameron in *Expression of Emotions* is likely because of her preference for sentimental, otherworldly staging and her painterly use of the medium—controversy repeatedly arose over whether her out-of-focus style indicated an artistic approach or just a dilettantish lack of skill. See Rosenblum 1984, 208–243. On the artistic value of the unfocused perspective in the history of art, see Ullrich 2002; on Cameron see Cox and Ford 2003.

35. On Cameron and Rejlander see Frizot 1998, 185–196; on Rejlander's allegory see Jones 1973, 17–23, and Brauchitsch 2002, 40f.

36. One photograph shows a smiling Rejlander with his famous picture of the crying baby and the comment that his wife had asked him to send the picture to convince Darwin that he was also capable of "more amiable expressions." See DAR 53.1.46.

37. Quoted in Desmond and Moore 1991, 594. On the use of heliotype printing in the book see Prodger 1998b, 156.

38. On miniature portraits meant for private use see Beyer 2003, 113f.

39. See Darwin 1872b, 258.

40. For Darwin's commentary on the scene see Darwin 1877, 233. For Dickens's quote see Charles Dickens, "Book Illustrations," in *All the Year Round*, August 10, 1867. Quoted in Frizot 1998, 189f.

41. On Duchenne's photographs in *Expression* see Darwin 1872b, Plate II, Figures 1–2; Plate III, Figures 4–6; 299, Figure 20; Plate VII, Figure 2; 306, Figure 21.

42. See Prodger 1998c, 404.

43. See Darwin 1872b, 14 and 203–204; on Duchenne see Dupouy 2003, 6.

44. See Darwin 1872b, 203–204.

45. See Darwin 1872b, 202. On the identity of the images see Prodger 1998a, 116.

46. See Darwin 1872b, 75–76f.

47. Quoted after Prodger 1998c, 405.

48. See Prodger 1998b.

49. See Prodger 1998b, 173.

50. Quoted in Prodger 1998b, 177. On ethnographical and anthropological photography and criticism of it, see Edwards 1990 and 1992, Wallis 1995, and Stepan 2002. The institution possessing Agassiz's

daguerreotypes no longer permits their reproduction; copies are included in Wallis 1995.

51. Wallis draws on Alan Sekula's essay "The Body and the Archive" in making this distinction between type and portrait. See Wallis 1995, 54. On the individual character of the portrait see Boehm 1985.

52. On this scientific criticism see Ekman's introduction to Darwin 1998 [1872]. Prodger, on the other hand, considers Darwin part of a scientific trend that Daston and Galison describe as the "ideal of mechanical objectivity." See Prodger 1998b, 177–181.

53. The practice of naming the animals observed in behavioral research was first reintroduced by Jane Goodall. On the history of behavioral research using primates see Haraway 1989.

54. Darwin ordered sixty thousand full-sized prints (nine inches by twelve inches) and two hundred and fifty thousand visiting cards (two inches by three inches). See Prodger 1998b, 173.

55. The first movement study of an animal was photographed by the American Edweard J. Muybridge, who captured images of a galloping race horse in 1879. See Frizot 1998, 243–257.

56. Unpublished letter from Charles Darwin to Briton Rivière, May 19, 1872, in DAR 147.320.

57. See Prodger 1998a, 95.

58. See Darwin 1872b, 15. Darwin here echoes Lessing's *Laocoon* essay.

59. See Darwin 1882, 1, and Woolner 1917, 21.

60. See Darwin 1872b, 52–53, Figures 5–6 (dog); 59, Figure 10 (cat); 136, Figures 16–17 (macaque); 141, Figure 18 (chimpanzee).

61. Guides for pet owners began appearing in the 1850s. See Kean 1998, 47ff. Rubin 2003 considers this issue in terms of art history. The most prominent example of the alienation theory is put forth in Berger 2003a. Also see Artinger 1995 and Kaselow 1999.

62. Kean names the city as the place where, in the nineteenth century, the most animals could be seen and where they aroused the most attention. See Kean 1998, 28ff. On the history of industrial, large-scale methods of keeping animals, see the 2002 exhibition catalog of the *Stiftung Deutsches Hygiene Museum*. Also see Wollschläger 2002 and Lackner 2004. On animals as a mirror of the English class system see Ritvo 1987. Animals bred by humans, especially pigeons, play a key role in both *Origin* and *Variation*. See Secord 1981 and 1985. On anthropomorphism in research see Mitchell 2005 and Serpell 2005. Bühler and Rieger 2006 examine the animal as a subject capable of cognition.

63. Unpublished letter from Harry Huxley to Charles Darwin, January 17, 1872. See DAR 166.2.

64. On Toby see Garland 1867; for the citation about Jerry see Blunt 1976, 96. Periodicals such as the popular *Penny Magazine*, established in 1832 as the first illustrated publication from the "Society for the Diffusion of Useful Knowledge," and with a circulation of up to two hundred thousand, featured an engraving of a zoo animal in every issue. See Kean 1998, 46. On animal shows in London see Altick 1978 and Goodall 2002.

65. In a discussion of the history of the depiction of dogs, William Secord indicates that portraits of the animal were a nineteenth-century phenomenon: "The mid- to late nineteenth century saw the emergence of three types of dog painting: the pet portrait, the sporting dog portrait and the pure-bred dog portrait." See Secord 2002, 12. On the social history of the portrait, especially the inflationary proliferation of the portrait in post-Revolutionary France, see Beyer 2003, 291f. On the ways in which the English social classes were mirrored in the animal kingdom see Ritvo 1987.

66. The organization became the Royal Society for the Prevention of Cruelty to Animals (RSPCA) under the patronage of Queen Victoria in 1840. On this topic as well as animal shelters, vegetarianism, and the anti-vivisection movement see the multifaceted and illuminating study of the history of the animal protection movement in Kean 1998. On vivisection see also Rupke 1987 and White 2005. Darwin's attitude toward vivisection remained ambivalent: on the one hand he defended the scientific value of the process, while on the other he stressed the fact that he had never dissected a living animal. See Browne 2002, 418–23 and 486. On Darwin's involvement on behalf of animal rights see White 2006, 110.

67. On the history of animal illustration see Nissen 1969 and 1978, Dance 1978, and Blum 1993.

68. Little is known about Thomas William Wood. As an example of his many drawings in the *Illustrated London News* see the antelope on February 25, 1865; on the argus pheasant in *The Field* see Tegetmeier 1871 and 1874. On *Bible Animals* see Wood 1869. On the drawings for Darwin see Prodger 1998a, 92–94.

69. For his work as a whole see Schulze-Hagen and Geus 2000. On the *Proceedings and Transactions* see Root and Johnson 1986a and 1986b. On the directory of his illustrations for travel books, etc., see the bibliography to Palmer 1895. For the commissions from Darwin see also Prodger 1998a, 95–97.

70. The quote is from William Michael Rossetti, the brother of Dante Gabriel, and reads in full: "And all the Pre-Raphaelites, including my brother, were delighted with his acute and minute observations,

and delicate precision of rendering." See Palmer 1895, 67. For the Landseer quote see Palmer 1895, 68.

71. On Polly and Huxley see Browne 2002, 361. On Darwin's discussion of Polly see Darwin 1872b, 43f.

72. See Armstrong 1891, 12.

73. On Rivière see Armstrong 1891. On his drawings for Darwin see Prodger 1998a, 98f. For their correspondence see DAR 176.174–182. On the dog in art see Clark 1977 and Secord 2002.

74. See Notebook N, 117.

75. See Darwin 1882, i, 103.

76. See Darwin 1875b, 146.

77. Unpublished correspondence. See DAR 176.174–182.

78. Unpublished letter from Briton Rivière to Charles Darwin, April 3, 1872, available in DAR 176.175. In this letter Rivière responds that he shares Darwin's criticism of the picture. The letter from Darwin with the criticism itself has not been preserved. On Landseer's painting see Ormond 1981, 197f. Darwin's archive contains a print of *Alexander and Diogenes* at DAR 53.1.141.

79. On grinning dogs see Darwin 1872b, 121; on the mention of Rivière see Darwin 1872b, 26.

80. On the iconography of the dog see Agthe 1987, Prange 2002, Secord 2002, and Rubin 2003.

81. See Goethe 1809, 171. The assistant with whom Ottilie had just discussed pedagogy also slams the book shut again as soon as he discovers it in the library, apparently because he considers it inappropriate reading for a lady. For the iconography of the ape see Janson 1952, Zimmerman 1991, Corbey and Theunissen 1995, Marret 2001, and Goodall 2002.

82. Quoted in Blunt 1976, 142f.

83. See Browne 2002, 450.

84. The report appeared in *Purchas His Pilgrimage*, quoted in Huxley 1863, 3–5.

85. See Schulze-Hagen and Geus 2000, 206.

86. The specimen entered the collection of the Paris museum on January 16, 1852, but the description was first published six years later. See Geoffroy Saint Hilaire 1858–1861. The extent to which a specimen in alcohol would decompose was recorded in a daguerreotype taken in the British Museum in 1858 after the vat was opened. See also Schulze-Hagen and Geus 2000, 196. The scientific description of this animal also first appeared several years later; see Owen 1865.

87. On the iconography of the "wild man" see Zapperi 2004, 24.

88. On Du Chaillu see Vaucaire 1930, Patterson 1974, Blunt 1976, 135–146, Mandelstam 1994, McCook 1996, and Browne 2002, 156–160.

89. See Gray 1861b, 695. Gray's accusations, which he first published on March 23 in the *Athenaeum*, unleashed a debate in the press that lasted until October. See Mandelstam 1994 and McCook 1996. In July the *Athenaeum* began to publish contributions to the debate under the heading "The Gorilla War." See *Athenaeum* no. 1755, 1861. A further arena for debate was the anatomical discussion of the gorilla as humanity's closest relative carried out between Owen and Huxley. See Browne 2002, 119f.

90. For the accusations, see Gray 1861b, 695. For Du Chaillu's defense, see Du Chaillu 1861c, 694. The frontispiece is not signed by Wolf but his biographer, Palmer, also ascribes it to him. See Palmer 1895, 156f.

91. On the quote and the essay in the *Proceedings* see Reade 1863, 212.

92. See Zimmerman 1991, 667.

93. See Young 1992, 23f., Browne 1995, 196ff., and Browne 2001, 500.

94. For the picture from the *Illustrated Police News* see Wood 2000, 191. On the history of slavery see Andrews 2004.

95. On the visual depiction of slavery see Wood 2000.

96. See Haeckel 1868, 555. Quoted after the original German. On the frontispiece see Gursch 1981 and especially Hopwood 2006, 280f. On Huxley see Huxley 1863b, 89, Figure 17.

97. On *Punch* see Browne 2001, 501; on the *Water-Babies* see Kingsley 1984, 266. *Alice in Wonderland* appeared in 1865, after Kingsley's work, which was first published as a serial in *Macmillan's Magazine* in 1862 and in book form in 1863. See Browne 2002, 160ff. The first edition from 1863 contained two illustrations by the painter Sir Joseph Noel Paton. The edition illustrated by the *Punch* cartoonist Linley Sambourne appeared in 1885.

98. Quoted in Browne 2002, 308.

99. On the quotation in the original see Figure 77. The American Society for the Prevention of Cruelty to Animals was founded in 1866 by Henry Bergh on the model of the English Royal Society for the Prevention of Cruelty to Animals, which had existed since 1824. On the caricature see Browne 2001.

100. In terms of evolutionary history, the reader learns only that Gratiolet considers the gorilla to be a "highly developed mandrill." See 1882, 177 and index. In *Expression of Emotions* see Darwin 1872b, 95–95 and 234. The first gorilla arrived at the London Zoo in 1887.

101. See Darwin 1882, 63–64.

102. *Descent of Man* appeared on February 24, 1871.

103. The incident is related in a letter that Darwin wrote to Joseph Wolf on March 3, 1871, to ask the artist to draw the macaque for him. It is reprinted in Alfred H. Palmer, *The Life of Joseph Wolf: Animal Painter* (London: Longmans, 1895), 193f.

104. See Palmer 1895, 194. On Darwin's visit, see Palmer 1895, 197f. On the macaque's enjoyment of being petted, see Darwin 1872b, 135.

105. See Spencer 1860. In this essay, Spencer formulates the thesis that Darwin further develops in *Expression of the Emotions:* that such expressions were originally due to an excess of nervous energy that has become habitual. See Darwin 1875b, 71. Spencer coined the expression "survival of the fittest" in his 1864 *Principles of Biology.* Darwin adopted it in 1872 in the sixth edition of *Origin of Species,* usually in the combination "natural selection; or the survival of the fittest"—a phrase which also forms the title of Chapter 4 in this edition.

106. On the laughing orangutans and chimpanzees, see Darwin 1872b, 132. The picture of the chimpanzee was drawn by Thomas Wood; see Darwin 1875b, 171, Figure 18.

107. On the image of the gorilla see Genge 2002, Lange 2006, and Voss 2009b. On the dominating topos of struggle in the reception of the theory of evolution see Ruse 1979, Kleeberg 2004, and Voss 2009a.

Conclusion

1. Thanks to Paul White for the opportunity to read his unpublished paper "Darwin and the Imperial Archive," which extensively considers Darwin's ethnological researches and materials.

2. Robert FitzRoy's book appeared in 1839 with the title *Narrative of the Surveying Voyages of H. M. Ships* Adventures *and* Beagle, *Between the Years 1826 and 1836.* On the construction and dissemination of images of humans in the nineteenth century, see Lange 2006.

3. The edition published by Harper and Brothers was identical. On the inquiry to Murray see *Correspondence* 3, 381.

4. Desmond and Morris 2009, xvii.

5. See Darwin 1882, 65 (lowest barbarians); 101 (baboon in Brehm); 619 (baboon and Tierra del Fuegian). These statements are analyzed in detail in Sternberger 1977, 102f. For the account in *Journal of Researches,* see Darwin 1845, 216.

6. On the history of the Tierra del Fuegians see Hazlewood 2003, 309. On the quotation see Darwin 1882, 182.

7. See Prodger 1998a, 29f.; 32f.; 101f.

8. See Darwin 1882, 167. On Darwinism in the political sphere see Paul 2003.

9. See Darwin 1882, 175.

10. See Darwin 1872, 50 (plants) and 109 (beetles).

11. The most influential creator of such violent visual worlds was the American artist Charles Knight, who painted murals for both the American Museum of Natural History in New York and the Field Museum in Chicago. His best-known book, *Life Through the Ages*, tellingly first appeared in 1946, shortly after the end of the Second World War. For images of natural selection in the popular press of the nineteenth century see Voss 2009a.

12. This discussion follows Shapin's "workshop-model of science." See Shapin 1989, 562f.

13. Following Star and Griesemer, it is possible to describe the Galápagos finches as "boundary objects." See Star and Griesemer 1989.

14. The concept of collage, taken from art history, seems appropriate for describing more than just the production of Darwin's images in a narrow sense. The relationship the word suggests between science and history characterizes many other types of scientific production, even when they do not take place in pictures. Describing the scientist as a "tinkerer" emphasizes feeling one's way in experimentation and creating theories at a synchronous level. The related concept of collage points to the diachronous dimension of science contained in material. On the scientist as a tinkerer, see Rheinberger 2003a, 33.

15. For Darwin's reception in the arts see Kort and Hollein 2009, Donald and Munro 2009, and Larson and Brauer 2009.

16. See Browne 2002, 65 and 477.

17. The passage most frequently cited as proof of Darwin's aestheticism comes at the end of *Origin of Species:* "It is interesting to contemplate a tangled bank, clothed with many plants of many kinds, with birds singing on the bushes, with various insects flitting about, and with worms crawling through the damp earth, and to reflect that these elaborately constructed forms, so different from each other, and dependent upon each other in so complex a manner, have all been produced by laws acting around us." See Darwin 1872, 429. For Darwin as a romantic see Kohn 1996.

18. Ethological research in the twentieth century focused almost exclusively on wild animals. One interesting exception is the essay by Kaminski, Call, and Fischer 2004.

19. On the theory that humans are biology's "evil apes" see Kleeberg 2004.

20. On approaches to biologically explaining lavish ornamentation in contemporary science, see Gadagkar 2003 and Voland and Grammer 2003. Menninghaus 2003, especially 138ff., discusses the use of aesthetic theories to explain beauty and places them in their philosophical and literary context. The proximity of Haeckel's ideas to those from natural theology is analyzed in Kleeberg 2005.

21. A uniform conception of instinct has been the subject of intense criticism in recent behavioral biology. As a way to differentiate the idea of instinct, phenomena of cultural evolution among animals are being investigated. See De Waal and Tyack 2003.

22. See "Is Biology an Autonomous Science?" in Mayr 1997, 8–23, and Mayr 1998.

23. See Mayr 1997, 9.

24. On determinism and essentialism in the nature-vs.-nurture debate, see Kleeberg and Walter 2001.

BIBLIOGRAPHY

Agamben, Giorgio. 2004. *The Open: Man and Animal.* Stanford: Stanford University Press.

Agassiz, Louis. 1833–1843. *Recherches sur les poissons fossiles.* Ouvrage couronné par la Société géologique de Londres. 5 vols. Neuchâtel.

———. 1874. "Evolution and Permanence of Type." *Atlantic* 33: 92–101.

———. 1962 [1859]. *Essay on classification.* Edited by Edward Lurie. Cambridge, Mass.: Belknap Press of Harvard University Press.

———, and A. A. Gould. 1848. *Principles of Zoology. Touching the Structure, Development, Distribution and Natural Arrangement of the Races of Animals, Living and Extinct; with Numerous Illustrations. For the Use of Schools and Colleges. Part I. Comparative Physiology.* Boston: Kendall and Lincoln.

Agthe, Marion. 1987. *Das Bild des Hundes in Albrecht Dürers Gesamtwerk. Darstellungen und Deutungsversuche.* Bochum: Studienverlag Brockmeyer.

Allen, David E. 1994. *The Naturalist in Britain: A Social History.* Second edition. Princeton, N.J.: Princeton University Press.

Alpers, Svetlana. 1983. *The Art of Describing: Dutch Art in the Seventeenth Century.* 2nd rev. ed. Chicago: University of Chicago Press.

Alter, Stephen G. 1999. *Darwinism and the Linguistic Image. Language, Race, and Natural Theology in the Nineteenth Century.* Baltimore: Johns Hopkins University Press.

Altick, Richard D. 1978. *The Shows of London*. Cambridge, Mass.: Belknap Press of Harvard University Press.

Anderson, Patricia. 1991. *The Printed Image and the Transformation of Popular Culture*. Oxford: Clarendon Press.

Andrews, George Reid. 2004. *Afro-Latin America, 1800–2000*. Oxford: Oxford University Press.

Arburg, Hans Georg von. 1999. "Johann Caspar Lavaters Physiognomik. Geschichte, Methodik, Wirkung." In Gerda Mraz and Uwe Schlögl (eds.). *Das Kunstkabinett des Johann Caspar Lavater* (exhibition catalog), 40–59. Vienna: Böhlau.

Argyll, George Douglas Campbell Duke of. 1867. *The Reign of Law*. London: Alexander Strahan.

Armstrong, Walter. 1891. *Briton Rivière: His Life and Work*. London: Art Journal Office.

Artinger, Kai. 1995. *Von der Tierbude zum Turm der blauen Pferde. Die künstlerische Wahrnehmung der wilden Tiere im Zeitalter der zoologischen Gärten*. Berlin: Reimer.

Babbage, Charles. 1830. *Reflections on the Decline of Science in England and Some of Its Causes*. London: B. Fellowes.

Bachmann-Medick, Doris. 2006. *Cultural Turns. Neuorientierungen in den Kulturwissenschaften*. Hamburg: Rowohlt.

Barrett, Paul H., P. J. Gautrey, S. Herbert, D. Kohn, and S. Smith (eds.). 1987. *Charles Darwin's Notebooks, 1836–1844. Geology, Transmutation of Species, Metaphysical Inquiries*. London: British Museum (Natural History) and Cambridge University Press.

Barry, Martin. 1836–1837. "On the Unity of Structure in the Animal Kingdom." *Edinburgh New Philosophical Journal* 22: 116–141, 354–364.

Barsanti, Giulio. 1992. *La scala, la mappa, l'albero*. Florence: Sansoni.

Bartram, Michael. 1985. *The Pre-Raphaelite Camera: Aspects of Victorian Photography*. London: Weidenfeld and Nicolson.

Baumunk, Michael, and Jürgen Rieß (eds.). 1994. *Darwin und Darwinismus. Eine Ausstellung zur Kultur-und Naturgeschichte* (exhibition catalog). Berlin: Akademie Verlag.

Baxmann, Inge, M. Franz, and W. Schäffner (eds.). 2000. *Das Laokoon-Paradigma. Zeichenregime im 18. Jahrhundert*. Berlin: Akademie Verlag.

Beatty, John. 1984. "Chance and Natural Selection." *Philosophy of Science* 51: 183–211.

Bedini, Silvio A. 1997. *The Pope's Elephant: An Elephant's Journey from Deep in India to the Heart of Rome*. Manchester: Carcanet.

Beer, Gillian. 1996. "'Authentic Tidings of Invisible Things':
Vision and the Invisible in the Later Nineteenth Century." In
Teresa Brennan and Martin Jay (eds.). *Vision in Context.*
Historical and Contemporary Perspectives on Sight, 83–98. New
York: Routledge.
———. 2000. *Darwin's Plots. Evolutionary Narrative in Darwin,
George Eliot and Nineteenth-Century Fiction.* Second edition.
Cambridge: Cambridge University Press.
Bell, Charles. 1847. *The Anatomy and Philosophy of Expression as
Connected with the Fine Arts.* Fourth edition. London: Murray.
Beretta, Marco (ed.). 2005. *From Private to Public. Natural Collec-
tions and Museums.* Sagamore Beach, Mass.: Science History
Publications.
Berger, John. 1980a. "Why Look at Animals?" In John Berger.
About Looking, 1–28. New York: Pantheon.
———. 1980b. "The Suit and the Photograph: On August Sander."
In John Berger. *About Looking*, 31–40. New York: Pantheon.
Bewick, Thomas. 1797–1804. *History of British Birds.* 2 vols.
Newcastle.
Beyer, Andreas. 2003. *Portraits: A History.* New York: H. N. Abrams.
Bischoff, Theodor Ludwig Wilhelm. 1845. *Entwicklungsgeschichte
des Hunde-Eies.* Braunschweig: Vieweg.
Blum, Ann Shelby. 1993. *Picturing Nature: American Nineteenth
Century Zoological Illustration.* Princeton, N.J.: Princeton
University Press.
Blunt, William. 1976. *The Ark in the Park: The Zoo in the Nineteenth
Century.* London: Hamilton.
Bodmer, George R. 1997. "The Technical Illustration of Thomas
Henry Huxley." In Alan P. Barr (ed.). *Thomas Henry Huxley's
Place in Science and Letters: Centenary Essays*, 277–295. Athens,
Ga.: Georgia University Press.
Boehm, Gottfried. 1985. *Bildnis und Individuum. Über den
Ursprung der Porträtmalerei in der italienischen Renaissance.*
Munich: Prestel.
Bosch-Abele, Susanne. 2000. *Opposition mit dem Zeichenstift:
1830–1835. La Caricature.* Gelsenkirchen: Arachne.
Bowler, Peter. 1977. "Darwinism and the Argument from Design:
Suggestions for a Reevalution." *Journal for the History of
Biology* 10: 29–43.
———. 1992. *The Non-Darwinian Revolution: Reinterpreting a
Historical Myth.* New edition. Baltimore: Johns Hopkins
University Press.

————. 1996. *Life's Splendid Drama: Evolutionary Theory and the Reconstruction of Life's Ancestry, 1860–1940*. Chicago: University of Chicago Press.

————. 2003. *Evolution: The History of an Idea*. Third edition. Berkeley, Calif.: University of California Press.

Bowman, Robert I. 1961. *Morphological Differentiation and Adaptation in the Galápagos Finches*. Berkeley: University of California Press.

Brackmann, Arnold C. 1980. *A Delicate Arrangement: The Strange Case of Charles Darwin and Alfred Russel Wallace*. New York: Times Book.

Brake, Laurel. 2001. *Print in Transition, 1850–1910: Studies in Media and Book History*. New York: Palgrave.

Brauchitsch, Boris von. 2002. *Kleine Geschichte der Fotografie*. Stuttgart: Reclam.

Bredekamp, Horst. 1995. *The Lure of Antiquity and the Cult of the Machine: the Kunstkammer and the Evolution of Nature, Art, and Technology*. Princeton: M. Wiener.

————. 2004a. "Darwins Evolutionsdiagramm oder: Brauchen Bilder Gedanken?" In Wolfram Hogrebe and Joachim Bromand (eds.). *Grenzen und Grenzüberschreitungen. xix. Deutscher Kongreß für Philosophie, Bonn, 23–27. September 2002. Vorträge und Kolloquien*, 863–877. Berlin: Akademie Verlag.

————. 2004b. "'Die wilde Üppigkeit der Natur': Stricklands Karten und Darwins Kreise der Arten." In Natascha Adamowsky and Peter Matussek (eds.). *Auslassungen. Leerstellen als Movens der Kulturwissenschaft*, 341–352. Würzburg: Königshausen and Neumann.

————. 2005. *Darwins Korallen. Frühe Evolutionsmodelle und die Tradition der Naturgeschichte*. Berlin: Wagenbach.

Brednow, Walter. 1969. *Von Lavater zu Darwin*. Berlin: Akademie Verlag.

Briggs, Asa. 1988. *Victorian Things*. London: Batsford.

Brink-Roby, Heather. 2008. "Natural Representation. Diagram and Text in Darwin's 'On the Origin of Species.'" *Victorian Studies* 51: 247–273.

Brontë, Charlotte. 2002 [1847]. *Jane Eyre*. New York: Dover Publications.

Browne, Janet. 1980. "Darwin's Botanical Arithmetic and the 'Principle of Divergence,' 1854–1858." *Journal of the History of Biology* 13: 53–89.

———. 1983. *The Secular Ark. Studies in the History of Biogeography.* New Haven: Yale University Press.

———. 1985a. "Darwin and the Face of Madness." In W. F. Bynum, Roy Porter, and Michael Shepherd (eds.). *The Anatomy of Madness. Essays in the History of Psychiatry. Volume 1. People and Ideas,* 151–165. London: Tavistock Publishers.

———. 1985b. "Darwin and the Expression of the Emotions." In David Kohn (ed.). *The Darwinian Heritage,* 307–326. Princeton, N.J.: Princeton University Press in association with Nova Pacifica.

———. 1995. *Charles Darwin: Voyaging.* Princeton, N.J.: Princeton University Press.

———. 2001. "Darwin in Caricature. A Study in the Popularization and Dissemination of Evolution." *Proceedings of the American Philosophical Society* 145: 496–509.

———. 2002. *Charles Darwin: The Power of Place.* New York: Alfred A. Knopf.

———. 2003. "Darwin as a Celebrity." *Science in Context* 16: 175–194.

———, and Sharon Messenger. 2003. "Julia Pastrana: The Bearded and Hairy Female." *Endeavour* 27: 155–159.

Büchmann, Georg. 1871. *Geflügelte Worte. Der Citatenschatz des Deutschen Volks.* Sixth revised and expanded edition. Berlin: Haude und Spener.

Bühler, Benjamin, and Stefan Rieger (eds.). 2006. *Vom Übertier. Ein Bestiarium des Wissens.* Frankfurt am Main: Suhrkamp.

Burchfield, Joe D. 1974. "Darwin and the Dilemma of Geological Time." *Isis* 65: 301–321.

Burke, Peter. 2001. *Eyewitnessing: The Use of Images as Historical Evidence.* Ithaca, N.Y.: Cornell University Press.

Burkhardt, Frederick H., Sydney Smith, et al. (eds.). 1983–2005. *The Correspondence of Charles Darwin.* Vols. 1–14 (1821–1866). Cambridge: Cambridge University Press.

Burkhardt, Richard W. 1985. "Darwin on Animal Behavior and Evolution." In David Kohn (ed.). *The Darwinian Heritage,* 327–365. Princeton, N.J.: Princeton University Press in association with Nova Pacifica.

Camerini, Jane R. 1996. "Wallace in the Field." *Osiris* 11: 44–45.

Campbell, Bruce, and Elizabeth Lack. 1985. *A Dictionary of Birds.* Published for the British Ornithologists' Union. Calton: Poyser.

Campbell, John Angus. 1989. "The Invisible Rhetorician. Charles Darwin's 'Third Party' Strategy." *Rhetorica* 7: 55–85.

Campe, Rüdiger, and Manfred Schneider (eds.). 1996. *Geschichten der Physiognomik. Text, Bild, Wissen.* Freiburg im Breisgau: Rombach.

Carpenter, William. 1841. *Principles of General and Comparative Physiology.* Second edition. London: John Churchill.

Chambers, Robert. 1844. *Vestiges of the Natural History of Creation.* London: Churchill.

Clair, Jean (ed.). 1993. *L'âme au corps. Arts et sciences 1793–1993* (exhibition catalog). Paris: Réunion des musées nationaux with Gallimard/Electa.

Clausberg, Karl. 1997. "Psychogenese und Historismus: Verworfene Leitbilder und übergangene Kontroversen." In Olaf Breidbach (ed.). *Natur der Ästhetik—Ästhetik der Natur,* 139–166. Wien: Springer Verlag.

Corbey, Raymond, and Bert Theunissen (eds.). 1995. *Ape, Man, Apeman: Changing Views Since 1600.* Evaluative proceedings of the symposium, Leiden, the Netherlands, 28 June–1 July 1993. Leiden: Leiden University.

Cox, Julian, and Colin Ford. 2003. *Julia Margaret Cameron: The Complete Photographs.* Los Angeles, Calif.: J. Paul Getty Museum.

Crary, Jonathan. 1990. *Techniques of the Observer: On Vision and Modernity in the Nineteenth Century.* Cambridge, Mass.: MIT Press.

———. 2002. *Suspensions of Perception: Attention, Spectacle, and Modern Culture.* Cambridge, Mass.: MIT Press.

Cuvier, Georges de. 1799–1805. *Leçons d'anatomie comparée.* 5 vol. Paris: Baudouin.

———. 1825. *Discourse on the Revolutionary Upheavals on the Surface of the Globe and on the Changes Which They Have Produced in the Animal Kingdom. Translated from the French, with illustrations and a glossary.* London: Whittaker, Treacher and Arnot.

Dance, S. Peter. 1978. *The Art of Natural History: Animal Illustrators and Their Work.* Woodstock, N.Y.: Overlook Press.

Darwin, Charles Robert. 1839. *Journal of Researches into the Geology and Natural History of the Various Countries Visited by H.M.S. Beagle.* London: Colburn.

———. 1842. *The Structure and Distribution of Coral Reefs. Part I of The Geology of the Voyage of the Beagle.* London: Smith and Elder.

————. 1845. *Journal of Researches into the Geology and Natural History of the Various Countries Visited by H.M.S. Beagle.* Second edition. London: Murray.

————. 1846. *Geological Observations of South America: Being the Third Part of the Geology of the Voyage of the* Beagle, *Under the Command of Capt. Fitzroy, R.N. During the Years 1832 to 1836.* London: Smith Elder.

————. 1859. *On the Origin of Species by Means of Natural Selection, or the Preservation of Favoured Races in the Struggle of Life.* London: Murray.

————. 1871. *The Descent of Man and Selection in Relation to Sex.* 2 vols. London: Murray.

————. 1872a. *The Expression of the Emotions in Man and Animals.* London: John Murray.

————. 1872b. *On the Origin of Species by Means of Natural Selection, or the Preservation of Favoured Races in the Struggle of Life.* Sixth edition. London: Murray.

————. 1875. *The Variation of Animals and Plants Under Domestication.* Second edition. London: John Murray.

————. 1877. "A Biographical Sketch of an Infant." *Mind: A Quarterly Review of Psychology and Philosophy* 2 (July): 285–294.

————. 1881. *The Formation of Vegetable Mould, Through the Action of Worms.* London: John Murray.

————. 1882. *The Descent of Man and Selection in Relation to Sex.* Second edition. London: Murray.

————. 1933. *Charles Darwin's Diary of the Voyage of the H.M.S. Beagle.* Edited with an introduction by Nora Barlow. Cambridge: Cambridge University Press.

————. 1958. *The Autobiography of Charles Darwin 1809–1882.* Edited by Nora Barlow. London: Collins.

————. 1963. "Darwin's Ornithological Notes." Edited and with an introduction, notes, and appendix by Nora Barlow. *Bulletin of the British Museum (Natural History)* 2: 201–278.

————. 1986 [1841]. "The Zoology of the Voyage of H.M.S. *Beagle,* Under the Command of Captain FitzRoy, During the Years 1832–1836." Vol. 3. Birds. In Paul H. Barrett and R. B. Freeman (eds.). *The Works of Charles Darwin.* Vol. 5. New York: New York University Press. 1998 [2000].

————, and Alfred Russel Wallace. 1958. *Evolution by Natural Selection.* Darwin's "Sketch" of 1842, his "Essay" of 1844, and the Darwin-Wallace Papers of 1858 "On the Tendency of Species to Form Varieties." With an introduction by Sir Francis Darwin and a foreword by Sir Gavin de Beer. Cambridge, Mass.: Cambridge University Press.

Darwin, Francis (ed.). 1903. *More Letters of Charles Darwin. A Record of His Work in a Series of Hitherto Unpublished Letters.* 2 vols. London: Murray.

———, (ed.). 1959. *The Life and Letters of Charles Darwin. Including an Autobiographical Chapter.* Foreword by George Gaylord Simpson, in two volumes. New York: Basic.

Daston, Lorraine. 2004. "Type Specimen and Scientific Memory." *Critical Inquiry* 31: 153–182.

———, and Katherine Park. 1998. *Wonders and the Order of Nature, 1150–1750.* New York: Zone.

———, and Peter Galison. 2002. "Das Bild der Objektivität." In Peter Geimer (ed.). *Ordnungen der Sichtbarkeit*, 29–99. Frankfurt am Main: Suhrkamp.

———, and Gregg Mitman. 2005. *Thinking with Animals: New Perspectives on Anthropomorphism.* New York: Columbia University Press.

De Beer, Gavin. 1964. *Charles Darwin. A Scientific Biography.* New York: Doubleday.

De Chadarevian, Soraya. 1996. "Laboratory Science Versus Country-House Experiments: The Controversy Between Julius Sachs and Charles Darwin." *The British Journal for the History of Science* 29: 17–41.

———, and Nick Hopwood. 2004. *Models. The Third Dimension of Science.* Stanford, Calif.: Stanford University Press.

Desmond, Adrian. 1982. *Archetypes and Ancestors. Palaeontology in Victorian London, 1850–1875.* London: Blond and Briggs.

———. 1985. "The Making of Institutional Zoology in London, 1822–1836." *History of Science* 23: 153–85, 223–50.

———. 1994. *Huxley: The Devil's Disciple.* London: Michael Joseph.

———, and James R. Moore. 1991. *Darwin.* London: Michael Joseph.

———, and James Moore. 2009. *Darwin's Sacred Cause. Race, Slavery and the Quest for Human Origins.* London: Allen Lane.

Dettelbach, Michael. 1996. "Humboldtian Science." In Nicholas Jardine, James A. Secord, and Emma Spary (eds.). *Cultures of Natural History*, 287–304. Cambridge: Cambridge: University Press.

De Waal, Frans, and Peter Tyack (eds.). 2003. *Animal Social Complexity. Intelligence, Culture, and Individualized Societies.* Cambridge, Mass.: Harvard University Press.

Dickens, Charles. 1851. "The Tresses of the Day Star." *Household Words* 65: 289–291.

Di Gregorio, Mario. 1982. "In Search of the Natural System. Problems of Zoological Classification in Victorian Britain." *History and Philosophy of the Life Sciences* 4: 225–254.

———. 1987. "Hugh Strickland (1811–53) on Affinities and Analogies. Or, the Case of the Missing Key." *Ideas and Production* 34: 35–50.

———. 2005. *From Here to Eternity. Ernst Haeckel and Scientific Faith*. Göttingen: Vandenhoeck and Ruprecht.

———, and Nick Gill (eds.). 1990. *Charles Darwin's Marginalia*. Vol. 1. New York: Garland.

Dobai, Johannes. 1974. *Die Kunstliteratur des Klassizismus und der Romantik in England*. Vol. 2. Bern: Benteli.

Dommann, Monika. 2004. "Vom Bild zum Wissen. Eine Bestandsaufnahme wissenschaftshistorischer Bildforschung." *Gesnerus* 61: 77–89.

Donald, Diana. 2007. *Picturing Animals in Britain, 1750–1850*. New Haven: Yale University Press.

Donald, Diana, and Jane Munro (eds.). 2009. *Endless Forms. Charles Darwin, Natural Science and the Visual Arts* (exhibition catalog). New Haven: Yale University Press.

Down House. The Home of Charles Darwin. 1998. Catalog published by English Heritage. London: English Heritage.

Du Chaillu, Paul Belloni. 1861a. "The Geographical Features and Natural History of a Hitherto Unexplored Region of Western Africa." *Proceedings of the Royal Geographical Society of London* 5: 108–113.

———. 1861b. *Explorations and Adventures in Equatorial Africa with Account of the Manners and Customs of the People, and of the Chace of the Gorilla, Crocodile, Leopard, Elephant, Hippopotamus, and Other Animals*. London: Murray.

———. 1861c. "The New Traveller's Tales." *Athenaeum* 1752: 694–695.

———. 1867. *A Journey to Ashango-Land, and the Further Penetration into Equatorial Africa*. London: Murray.

Duden, Barbara, et al. (eds.). 2002. *Geschichte des Ungeborenen. Zur Erfahrungs- und Wissenschaftsgeschichte der Schwangerschaft, 17.–20. Jahrhundert*. Second edition. Göttingen: Vandenhoeck und Ruprecht.

Dupouy, Stéphanie. 2005. "Künstliche Gesichter: Rodolphe Töpffer und Duchenne de Boulogne." In Andreas Mayer and Alexandre

Métraux (eds.). *Kunstmaschinen. Spielräume des Sehens zwischen Wissenschaft und Ästhetik*, 24–60. Frankfurt am Main: Fischer.

Ebert-Schifferer, Sybille (ed.). 2002. *Deceptions and Illusions: Five Centuries of Trompe l'oeil Painting* (exhibition catalog). Washington: National Gallery of Art.

Ecker, Alexander Anton Pius. 1851–1859. *Icones Physiologicae: Erläuterungstafeln zur Physiologie und Entwicklungsgeschichte.* Leipzig: Voss.

Edwards, Elizabeth. 1990. "'Photographic Types': The Pursuit of a Method." *Visual Anthropology* 3: 235–258.

———, (ed.). 1992. *Anthropology and Photography 1860–1920*. New Haven: Yale University Press in association with the Royal Anthropological Society.

Elkins, James. 1999. *The Domain of Images*. Ithaca, N.Y.: Cornell University Press.

Ellegard, Alvor. 1990. *Darwin and the General Reader: The Reception of Darwin's Theory of Evolution in the British Periodical Press, 1859–1872*. With a new foreword by David Hull. Chicago: University of Chicago Press.

Engels, Eve-Marie (ed.). 1995. *Die Rezeption der Evolutionstheorien im 19. Jahrhundert*. Frankfurt am Main: Suhrkamp.

Farber, Paul Lawrence. 1976. "The Type-Concept in Zoology During the First Half of the Nineteenth Century." *Journal of the History of Biology* 9: 93–119.

———. 1977. "The Development of Taxidermy and the History of Ornithology." *Isis* 68: 550–566.

———. 1985. "Aspiring Naturalists and Their Frustrations: The Case of William Swainson (1789–1855)." In Alwyne Wheeler and James H. Price (eds.). *From Linneaus to Darwin. Commentaries on the History of Biology and Geology.* Papers from the fifth Easter meeting of the Society for the History of Natural History, March 28–31, 1983, 51–59. London: Society for the History of Natural History.

Foucault, Michel. 1971. *The Order of Things: An Archeology of the Human Sciences.* New York: Pantheon.

Fraassen, Bas C. van. 1980. *The Scientific Image.* Oxford: Clarendon Press.

Freeman, Richard Broke. 1978. "Darwin's Negro Bird-Stuffer." *Notes and Records of the Royal Society of London* 33: 83–86.

Frizot, Michel (ed.). 1998. *A New History of Photography.* Cologne: Könemann.

Gadagkar, R. 2003. "Is the Peacock Merely Beautiful or Also Honest?" *Current Science* 85: 1012–1020.

Galton, Francis. 1869. *Hereditary Genius: An Inquiry into Its Laws and Consequences.* London: Macmillan.

Garland, H. 1867. "Toby Is Wanted." *London Illustrated News* 28 (December 1867): 703–705.

Gasman, Daniel. 1971. *The Scientific Origins of National Socialism: Social Darwinism in Ernst Haeckel and the German Monist League.* London: Macdonald.

Geimer, Peter. 2003. "Weniger Schönheit. Mehr Unordnung: Eine Zwischenbemerkung zu 'Wissenschaft und Kunst.'" *Neue Rundschau* 114: 26–38.

Genge, Gabriele. 2002. "King Kong und die weiße Frau: Die Geschichte des Menschen im Bild des Affe." In Hans Körner and Angela Stercken (eds.). *Kunst, Sport und Körper,* 48–67. Ostfildern-Ruit: Hatje Cantz.

Geoffroy Saint-Hilaire, Isidore de. 1847. "Description des mammifères nouveaux ou imparfaitement connus de la collection du muséum d'histoire naturelle et remarques sur la classification et les caractères des mammifères. Quatrième mémoire, famille des singes, second supplément." *Archives du Muséum d'Histoire Naturelle* 10: 1–100 (Plates I–VI).

Gillespie, Neal C. 1977. "The Duke of Argyll, Evolutionary Anthropology, and the Art of Scientific Controversy." *Isis* 68: 40–54.

———. 1990. "Divine Design and the Industrial Revolution: William Paley's Abortive Reform of Natural Theology." *Isis* 81: 214–229.

Ginzburg, Carlo. 1983. *Spurensicherung. Die Wissenschaft auf der Suche nach sich selbst.* Berlin: Wagenbach.

Glick, Thomas F. (ed.). 1988. *The Comparative Reception of Darwinism.* Reprinted with a new preface. Chicago: University of Chicago Press.

Goethe, Johann Wolfgang von. 1809. "Elective Affinities." In Johann Wolfgang von Goethe: *Novels and Tales by Goethe.* Translated from the German by James Anthony Froude and R. Dillon Boylan, 1–245. London: G. Bell and Sons, 1911.

Gombrich, Ernst H. 1972. "Aims and Limits of Iconology." In *Symbolic Images: Studies in the Art of the Renaissance,* 1–25. Vol. 2. London: Phaidon.

———. 2002. *Art and Illusion: A Study in the Psychology of Pictorial Representation.* Sixth edition. New York: Phaidon.

Goodall, Jane R. 2002. *Performance and Evolution in the Age of Darwin. Out of the Natural Order*. London: Routledge.

Goritschnig, Ingrid. 1999. "Lavaters auserwählter Künstlerkreis." In Gerda Mraz and Uwe Schlögl (eds.). *Das Kunstkabinett des Johann Caspar Lavater* (exhibition catalog), 96–109. Vienna: Böhlau.

Gould, John. 1837a. "Remarks on a Group of Ground Finches from Mr. Darwin's Collection, with Characters of the New Species." *Proceedings of the Zoological Society* 5: 4.

———. 1837b. "Observations on the Raptorial Birds in Mr. Darwin's Collection, with Characters of the New Species." *Proceedings of the Zoological Society* 5: 9.

———. 1837c. "Exhibition of the Fissirostral Birds from Mr. Darwin's Collection, and Characters of the New Species." *Proceedings of the Zoological Society* 5: 22.

———. 1837d. "Exhibition of Three Species of the Genus Orpheus from Mr. Darwin's Collection, and Characters of the New Species." *Proceedings of the Zoological Society* 5: 27.

———. 1837e. "On a New Rhea (*Rhea Darwinii*) from Mr. Darwin's Collection." *Proceedings of the Zoological Society* 5: 35.

———. 1849–1861. *Monograph of Trochilidae, Or, Family of Humming-Birds*. 5 vols. London: Mintern Brothers.

———. 1861. "An Introduction to the Trochilidae; Or, Family of Humming-Birds." *Athenaeum*, no. 1774: 536–537.

Gould, Stephen Jay. 1985. *Ontogeny and Phylogeny*. Thirteenth edition. Cambridge, Mass.: Belknap.

———. 1987. "Bushes All the Way Down." *Natural History Magazine* 96: 12–19.

———. 1989. *Wonderful Life. The Burgess Shale and the Nature of History*. New York: Norton.

———. 1995. "Ladders and Cones: Constraining Evolution by Canonical Icons." In Robert B. Silvers (ed.). *Hidden Histories of Science*. New York: New York Review of Books, 37–67.

———, N. L. Gilinsky, and R. Z. German. 1987. "Asymmetry of Lineages and the Directions of Evolutionary Time." *Science* 236: 1437–1441.

Graham, John. 1979. *Lavater's Essays on Physiognomy. A Study in the History of Ideas*. Bern: Peter Lang.

Grandville, J. J. *Karikatur und Zeichnung. Ein Visionär der französischen Romantik* (exhibition catalog). 2000. Ostfildern-Ruit: Hatje Cantz.

Gray, John Edward. 1861a. "The New Traveller's Tales." *Athenaeum,* no. 1751: 662–663.

———. 1861b. "The New Traveller's Tales." *Athenaeum,* no. 1752: 695.

Grote, Andreas (ed.). 1994. *Macrocosmos in Microcosmos. Die Welt in der Stube. Zur Geschichte des Sammelns 1450–1800.* Opladen: Leske and Budrich.

Gruber, Howard E. 1974. *Darwin on Man. A Psychological Study of Creativity.* London: Wilwood House.

———. 1988. "Darwin's 'Tree of Nature' and Other Images of Wide Scope." In Judith Wechsler (ed.). *On Aesthetics in Science,* 121–140. Third edition. Boston: Birkhäuser.

Gunther, Albert E. 1972. "The Original Drawings of George Henry Ford." *Journal of the Society for the Bibliography of Natural History* 6: 139–142.

———. 1975a. *A Century of Zoology at the British Museum Through the Lives of Two Keepers, 1815–1914.* London: Dawsons of Pall Mall.

———. 1975b. "The Darwin Letters at Shrewsbury School." *Notes and Records of the Royal Society of London* 30: 25–43.

Gursch, Reinhard. 1981. *Die Illustrationen Ernst Haeckels zur Abstammungs- und Entwicklungsgeschichte.* Frankfurt am Main: P. D. Lang.

Haeckel, Ernst. 1866. *Generelle Morphologie der Organismen. Allgemeine Grundzüge der organischen Formen-Wissenschaft. Mechanisch begründet durch die von Charles Darwin reformirte Descendenz-Theorie.* Berlin: Reimer.

———. 1868. *Natürliche Schöpfungsgeschichte: Gemeinverständliche wissenschaftliche Vorträge über die Entwicklungsgeschichte im Allgemeinen und diejenige von Darwin, Goethe und Lamarck im Besonderen, über die Anwendung derselben auf den Ursprung des Menschen und andere damit zusammenhängende Grundfragen der Naturwissenschaft.* Berlin: Reimer.

———. 1874. *Anthropogenie oder Entwicklungsgeschichte des Menschen. Gemeinverständliche wissenschaftliche Vorträge über die Grundzüge der menschlichen Keimes- und Stammes-Geschichte.* Leipzig: Engelmann.

———. 1998. *Art Forms in Nature: The Prints of Ernst Haeckel. One Hundred Color Plates. With contributions from Olaf Breidbach.* New York: Prestel.

Hagner, Michael. 1999. "Enlightened Monsters." In William Clark, Jan Golinski, and Simon Schaffer (eds.). *The Sciences and*

Enlightened Europe, 175–217. Chicago: University of Chicago Press.

———. (ed.). 2001. *Ansichten der Wissenschaftsgeschichte*. Frankfurt am Main: Fischer.

———. 2004. *Geniale Gehirne. Zur Geschichte der Elitegehirnforschung*. Göttingen: Wallstein.

Haraway, Donna. 1989. *Primate Visions: Gender, Race, and Nature in the World of Modern Science*. New York: Routledge.

Hartley, Lucy. 2001. *Physiognomy and the Meaning of Expression in Nineteenth Century Culture*. Cambridge: Cambridge University Press.

Hazlewood, Nick. 2001. *Savage: The Life and Times of Jemmy Button*. New York: Thomas Dunne/St. Martin's.

Heesen, Anke te (ed.). 2002. *Cut and paste um 1900. Der Zeitungsausschnitt in den Wissenschaften*. Kaleidoskopien 4. Berlin: Vice Versa.

———, and Emma C. Spary (eds.). 2001. *Sammeln als Wissen. Das Sammeln und seine wissenschaftliche Bedeutung*. Göttingen: Wallstein.

Heintz, Bettina, and Jörg Huber (eds.). 2001. *Mit dem Auge denken. Strategien der Sichtbarmachung in wissenschaftlichen und virtuellen Welten*. Vienna: Springer.

Helmholtz, Hermann von. 1871. *Populäre wissenschaftliche Vorträge*. Vol. 2. Braunschweig: Vieweg.

Heppel, D. 1981. "The Evolution of the Code of Zoological Nomenclature." In Alwyne Wheeler and James H. Price (eds.). *History in the Service of Systematics*, 135–141. London: Society for the Bibliography of Natural History.

Herbert, Sandra (ed.). 1980. "The Red Notebook of Charles Darwin." *Bulletin of the British Museum (Natural History), Historical Series* 7: 1–164.

———. 1985. "Darwin the Young Geologist." In David Kohn (ed.). *The Darwinian Heritage*, 483–510. Princeton, N.J.: Princeton University Press in association with Nova Pacifica.

———. 1991. "Charles Darwin as a Prospective Geological Author." *The British Journal for the History of Science* 24: 159–192.

———. 2005. *Charles Darwin, Geologist*. New York: Cornell University Press.

Herschel, John Frederick William. 1831. *A Preliminary Discourse on the Study of Natural Philosophy*. London: Longman.

Hodge, M. Jonathan S. 1985. "Darwin as a Lifelong Generation Theorist." In David Kohn (ed.). *The Darwinian Heritage*,

207–243. Princeton, N.J.: Princeton University Press in association with Nova Pacifica.

———. 2003. "The Notebook Programmes and Projects of Darwin's London Years." In Jonathan Hodge and Gregory Radick (eds.). *The Cambridge Companion to Darwin*, 40–68. Cambridge: Cambridge University Press.

Hopwood, Nick. 2000. "Producing Development: The Anatomy of Human Embryos and the Norms of Wilhelm His." *Bulletin of the History of Medicine* 74: 29–79.

———. 2002a. *Embryos in Wax. Models from the Ziegler Studios.* Cambridge: Whipple Museum of the History of Science.

———. 2002b. "Embryonen 'auf dem Altar der Wissenschaft zu opfern': Entwicklungsreihen im späten neunzehnten Jahrhundert." In Barbara Duden et al. (eds.). *Geschichte des Ungeborenen. Zur Erfahrungs- und Wissenschaftsgeschichte der Schwangerschaft, 17.–20. Jahrhundert*, 237–272. Second edition. Vandenhoeck and Ruprecht.

———. 2005. "Visual Standards and Disciplinary Change: Normal Plates, Tables and Stages in Embryology. *History of Science* 43: 239–303.

———. 2006. "Pictures of Evolution and Charges of Fraud: Ernst Haeckel's Embryological Illustrations." *Isis* 97: 260–301.

Hoyningen-Huene, Paul. 1987. "On the Distinction Between the 'Context' of Discovery and the 'Context' of Justification." *Epistemologia* 10: 81–88.

Huxley, Thomas Henry. 1863. *Evidence as to Man's Place in Nature.* London: Williams and Norgate.

Ivins, William. 1953. *Prints and Visual Communication.* London: Routledge and Kegan Paul.

Jackson, Christine E., and Maureen Lambourne. 1990. "Bayfield: John Gould's Unknown Colourer." *Archives of Natural History* 17: 189–200.

Jahn, Ilse (ed.). 1998. *Geschichte der Biologie. Theorien, Methoden, Institutionen, Kurzbiographien.* Third revised and expanded edition. Jena: Fischer.

Janson, Horst W. 1952. *Apes and Ape Lore in the Middle Ages and the Renaissance.* London: Warburg Institute.

Jardine, Nicholas, James A. Secord, and Emma Spary (eds.). 1996. *Cultures of Natural History.* Cambridge: Cambridge University Press.

Jardine, William S. 1858. *Memoirs of Hugh Edwin Strickland.* London: Voorst.

Jones, Caroline, and Peter Galison (eds.). 1998. *Picturing Science, Producing Art*. New York: Routledge.

Jones, Edgar Y. 1973. *Father of Art Photography: O. G. Rejlander 1813–75*. Newton Abbot: David and Charles.

Jones, Greta. 2002. "Alfred Russel Wallace, Robert Owen and the Theory of Natural Selection." *British Journal for the History of Science* 35: 73–96.

Junker, Thomas, and Uwe Hoßfeld. 2001. *Die Entdeckung der Evolutionstheorie. Eine revolutionäre Theorie und ihre Geschichte*. Darmstadt: Wissenschaftliche Buchgesellschaft.

Junker, Thomas, and Marsha Richmond (ed.). 1996. *Charles Darwins Briefwechsel mit deutschen Naturforschern. Ein Kalendarium mit Inhaltsangaben, biographischem Register und Bibliographie*. Marburg an der Lahn: Basilisken Presse.

Kaminski, J., J. Call, and J. Fischer. 2004. "Word Learning in a Domestic Dog. Evidence for 'Fast Mapping.'" *Science* 304: 1682–1683.

Kaselow, Gerhild. 1999. *Die Schaulust am exotischen Tier. Studien zur Darstellung des zoologischen Gartens in der Malerei des 19. und 20. Jahrhunderts*. Hildesheim: Georg Olms.

Kean, Hilda. 1998. *Animal Rights: Political and Social Change in Britain Since 1800*. London: Reaktion.

Kemp, Martin. 1981. *Leonardo da Vinci. The Marvellous Works of Nature and Man*. London: Dent.

———. 2000. *Visualizations: The Nature Book of Art and Science*. Berkeley, Calif.: University of California Press.

Kemp, Wolfgang. 1979. *". . . einen wahrhaft bildenden Zeichenunterricht überall einzuführen." Zeichnen und Zeichenunterricht der Laien 1500–1870*. Frankfurt am Main: Syndikat.

Kerr, David S. 2000. *Caricature and French Political Culture, 1830–1848: Charles Philipon and the Illustrated Press*. Oxford: Clarendon.

Keynes, Richard D. (ed.). 1988. *Charles Darwin's* Beagle *Diary. Edited by Richard Keynes*. Cambridge: Cambridge University Press.

———. 1997. "Steps on the Path to the *Origin of Species*." *Journal of Theoretical Biology* 187: 461–471.

———. 2003. *Fossils, Finches and Fuegians: Charles Darwin's Adventures and Discoveries on the Beagle*. London: Harper Collins.

Kimler, William C. 1983. "One Hundred Years of Mimicry. History of an Evolutionary Exemplar." PhD diss., University of Michigan.

King-Hele, Desmond. 2004. Prologue. "Catching Up with Erasmus Darwin in the New Century," 13–29. In Christopher U. M. Smith and Robert Arnott (eds.). *The Genius of Erasmus Darwin.* Aldershot: Ashgate.

Kingsley, Charles. 1984 [1885]. *The Water-Babies: A Fairy Tale for a Land-Baby.* New edition with one hundred illustrations by Linley Sambourne. Facsimile edition. London: Chancellor.

Kleeberg, Bernhard. 2004. "Die vitale Kraft der Aggression. Evolutionistische Theorien des bösen Affen 'Mensch.'" In Ulrich Bröckling, Benjamin Bühler, et al. (eds.). *Disziplinen des Lebens. Zwischen Anthropologie, Literatur und Politik,* 203–222. Tübingen: Gunter Narr.

―――. 2005. *Theophysis. Ernst Haeckels Philosophie des Naturganzen.* Cologne: Böhlau.

―――, and Tilmann Walter. 2001. "Der mehrdimensionale Mensch. Zum Verhältnis von Biologie und kultureller Entwicklung." In Bernhard Kleeberg, Stefan Metzger, Wolfgang Rapp, and Tilmann Walter (eds.). *Die List der Gene. Strategeme eines neuen Menschen,* 21–67. Tübingen: Gunter Narr.

Kockerbeck, Christoph. 1997. *Die Schönheit des Lebendigen. Ästhetische Naturwahrnehmung im 19. Jahrhundert.* Vienna: Böhlau.

Koebner, Thomas (ed.). 1989. *Laokoon und kein Ende. Der Wettstreit der Künste.* Munich: Edition Text and Kritik.

Kohn, David. 1980. "Theories to Work by: Rejected Theories, Reproduction, and Darwin's Path to Natural Selection." *Studies in the History of Biology* 4: 67–170.

―――, (ed.). 1985. *The Darwinian Heritage.* Princeton: Princeton University Press in association with Nova Pacifica.

―――. 1996. "The Aesthetic Construction of Darwin's Theory." In Alfred I. Tauber (ed.). *The Elusive Synthesis. Aesthetics and Science,* 13–48. Dordrecht: Kluwer Academic Publishers.

Kort, Pamela, and Max Hollein (eds.). 2009. *Darwin and the Search for Origins* (exhibition catalog). Frankfurt a. M.: Wienand Verlag.

Kottler, Malcolm J. 1978. "Charles Darwin's Biological Species Concept and Theory of Geographic Speciation: The Transmutation Notebooks." *Annals of Science* 35: 275–298.

Krasner, James. 1992. *The Entangled Eye. Visual Perception and the Representation of Nature in Post-Darwinian Narrative.* New York: Oxford University Press.

Krauße, Erika. 1995. "Haeckel. Promorphologie und 'evolutionistische' ästhetische Theorie. Konzept und Wirkung." In Eve-Marie

Engels (ed.). *Die Rezeption von Evolutionstheorien im 19. Jahrhundert*, 347–394. Frankfurt am Main: Suhrkamp.

Lack, David. 1947. *Darwin's Finches*. Cambridge: Cambridge University Press.

Lackner, Helmut. 2004. "Ein 'blutiges Geschäft.' Kommunale Vieh-und Schlachthöfe im Urbanisierungsprozeß des 19. Jahrhunderts. Ein Beitrag zur Geschichte der städtischen Infrastruktur." *Technikgeschichte* 71: 89–138.

Lam, H. J. 1936. "Phylogenetic Symbols, Past and Present." *Acta Biotheoretica* 2: 153–194.

Lambourne, Maureen. 1987a. "John Ruskin, John Gould and Ornithological Art." *Apollo* 126: 185–189.

———. 1987b. *John Gould: Bird Man*. Milton Keynes: Osberton Productions.

———. 1990. *The Art of Bird Illustration*. Secaucus, N.J.: Wellfleet.

Lange, Britta. 2006. *Echt, unecht, lebensecht. Menschenbilder im Umflauf*. Berlin: Kadmos Kulturverlag.

Larson, Barbara, and Fae Brauer (eds.). 2009. *The Art of Evolution: Darwin, Darwinism, and Visual Culture*. Lebanon: Dartmouth College Press.

Larson, James L. 1994. *Interpreting Nature: The Science of Living Form from Linnaeus to Kant*. Baltimore: Johns Hopkins University Press.

Latour, Bruno. 1990. "Drawing Things Together." In Michael Lynch and Steve Woolgar (eds.). *Representation in Scientific Practice*, 19–68. Cambridge, Mass.: MIT Press.

———. 1997. "Der Pedologenfaden von Boa Vista. Eine photo-philosophische Montage." In Hans-Jörg Rheinberger, Michael Hagner, and Bettina Wahrig-Schmidt (eds.). *Räume des Wissens. Repräsentation, Codierung, Spur*, 213–263. Berlin: Akademie Verlag.

———, and Peter Weibel (eds.). 2002. *Iconoclash: Beyond the Image Wars in Science, Religion and Art*. Cambridge, Mass.: MIT Press.

Lechtreck, Hans-Jürgen. 2008. "Evolution vor der Kamera. Roger Fenton und Richard Owen im British Museum, 1956–1958." *Fotogeschichte. Beiträge zur Geschichte und Ästhetik der Fotografie* 28: 39–56.

Lefèvre, Wolfgang. 1984. *Die Entstehung der biologischen Evolutionstheorie*. Vienna: Ullstein.

Lenoir, Timothy. 1989. *The Strategy of Life: Teleology and Mechanics in Nineteenth-Century German Biology*. Chicago and London: University of Chicago Press.

————, (ed.). 1998. *Inscribing Science. Materiality and Text in Scientific Communication.* Stanford, Calif.: Stanford University Press.

Lepenies, Wolf. 1976. *Das Ende der Naturgeschichte. Wandel kultureller Selbstverständlichkeiten in den Wissenschaften des 18. und 19. Jahrhunderts.* Munich: Hanser.

Lightman, Bernhard (ed.) 1997. *Victorian Science in Context.* Chicago: University of Chicago Press.

————. 2000. "The Visual Theology of Victorian Popularizers of Science from Reverent Eye to Chemical Retina." *Isis* 91: 651–680.

Lovejoy, Arthur O. 1964. *The Great Chain of Being: A Study of the History of an Idea.* Cambridge, Mass.: Harvard University Press.

Lyell, Charles. 1990–1991 [1830–1833]. *Principles of Geology, Being an Attempt to Explain the Former Changes of the Earth's Surface, by Reference to Causes Now in Operation.* 3 vols. Facsimile of the first edition. Chicago: University of Chicago Press.

Lyons, Sherrie. 1997. "Convincing Men They Are Monkeys." In Alan Barr (ed.). *Thomas Henry Huxley's Place in Science and Letters. Centenary Essays,* 95–118. Athens, Ga.: University of Georgia Press.

Maar, Christa, and Hubert Burda (eds.). 2004. *Iconic Turn. Die neue Macht der Bilder.* Cologne: Dumont.

Macho, Thomas. 1997. "Der Aufstand der Haustiere." In Marina Fischer-Kowalski et al. (eds.). *Gesellschaftlicher Stoffwechsel und Kolonisierung von Natur. Ein Versuch in Sozialer Ökologie,* 177–200. Amsterdam: Gordon and Breach Verlag Facultas.

MacLeay, William Sharp. 1819–1821. *Horae Entomologicae, or Essays on the Annulose Animals.* London: Bagster.

Mandelstam, Joel. 1994. "Du Chaillu's Stuffed Gorillas and the Savants from the British Museum." *Notes and Records of the Royal Society of London* 48: 227–245.

Mandelstam, Osip. 1991. "Rund um die Naturforscher." In *Gespräche über Dante. Gesammelte Essays II, 1925–1935.* Zurich: Ammann.

Marret, Bertrand. 2001. *Portraits de l'artiste en singe. Les singerie dans la peinture.* Paris: Somogy Éditions d'Art.

Mayr, Ernst. [1942] 1982. *Systematics and the Origin of Species: With an Introduction by Niles Eldredge.* New York: Columbia University Press.

————. 1992. "Darwin's Principle of Divergence." *Journal of the History of Biology* 25: 343–359.

————. 1996. "What Is a Species, and What Not?" *Philosophy of Science* 63: 262–277.

————. 1997. *This Is Biology: The Science of the Living World.* Cambridge, Mass.: Belknap Press of Harvard University Press.

————. 1998. *Philosophie der Biologie.* Berichte und Abhandlungen der Berlin-Brandenburgischen Akademie der Wissenschaften. Vol. 5. Berlin: Akademie Verlag.

————. 2001. *What Evolution Is.* New York: Basic.

Mazzolini, Renato G. (ed.). 1993. *Non-Verbal Communication in Science Prior to 1900.* Florence: Olschki.

McCook, Stuart. 1996. "'It May Be Truth, but It Is Not Evidence': Paul du Chaillu and the Legitimation of Evidence in the Field Sciences." *Osiris* 11: 177–197.

McEvey, Allan. 1973. *John Gould's Contribution to British Art: A Note on Its Authenticity.* Sydney: Sydney University Press.

McOuat, Gordon. 1996. "Species, Rules, and Meaning: The Politics of Language and the Ends of Definitions in Nineteenth-Century Natural History." *Studies in the History and Philosophy of Science* 27: 473–519.

————. 2001. "Cataloguing Power: Delineating 'Competent Naturalists' and the Meaning of Species in the British Museum." *British Journal for the History of Science* 34: 1–28.

Meijer, Miriam Claude. 1999. *Race and Aesthetics in the Anthropology of Petrus Camper.* Amsterdam: Rodopi.

Menninghaus, Winfried. 2003. *Das Versprechen der Schönheit.* Frankfurt am Main: Suhrkamp.

Mitchell, Sandra M. 2005. "Anthropomorphism and Cross-Species Modeling." In Lorraine Daston and Gregg Mitman (eds.). *Thinking with Animals. New Perspectives on Anthropomorphism,* 100–117. New York: Columbia University Press.

Mitchell, Timothy. 1989. "The World as Exhibition." *Comparative Studies in Society and History* 31: 217–236.

Mitchell, W. J. Thomas. 1998. *The Last Dinosaur Book.* Chicago: The University of Chicago Press.

Mivart, St. George Jackson. 1871. *On the Genesis of Species.* Second edition. London: Macmillan.

————. 1873. "On Lepilemur and Cheirogaleus, and the Zoological Rank of the Lemuroidea." *Proceedings of the Zoological Society* 41: 485–510.

Moore, James R. 1985. "Darwin of Down: The Evolutionist as Squarson-Naturalist." In David Kohn (ed.). *The Darwinian*

Heritage, 435–482. Princeton, N.J.: Princeton University Press in Association with Nova Pacifica.

Mraz, Gerda, and Uwe Schlögl (eds.). 1999. *Das Kunstkabinett des Johann Caspar Lavater* (exhibition catalog). Vienna: Böhlau.

Müller-Wille, Staffan. 1999. *Botanik und weltweiter Handel. Zur Begründung eines natürlichen Systems der Pflanzen durch Carl von Linné (1707–78)*. Berlin: Verlag für wissenschaftliche Bildung.

———. 2001. "Gardens of Paradise." *Endeavour* 25: 49–54.

———. 2002. "Text, Bild und Diagramm in der klassischen Naturgeschichte." *Kunsttexte.de* 4: 1–14.

———. 2005. "Konstellation, Serie, Formation. Genealogische Denkfiguren bei Harvey, Linnaeus und Darwin." In Sigrid Weigel, Ohad Parnes, Ulrike Vedder, and Stefan Willer (eds.). *Generation: Zur Genealogie des Konzepts, Konzepte von Genealogie*, 215–233. Paderborn: Wilhelm Fink Verlag.

Munro, Jane. 2009. "'More Like a Work of Art Than of Nature': Darwin, Beauty and Sexual Selection." In Diana Donald and Jane Munro (eds.). *Endless Forms. Charles Darwin, Natural Science and the Visual Arts* (exhibition catalog), 253–291. New Haven: Yale University Press.

Nissen, Carl. 1969–1978. *Die zoologische Buchillustration, ihre Bibliographie und Geschichte*. 2 vols. Stuttgart: Hiersemann.

Nochlin, Linda. 2003. "The Darwin Effect. Evolution and Nineteenth-Century Visual Culture." In http://www.19thc-artworldwide.org/spring_03/index.html.

Numbers, Ronald L. (ed.). 1999. *Disseminating Darwinism: The Role of Place, Race, Religion, and Gender*. Cambridge: Cambridge University Press.

O'Hara, Robert J. 1988. "Diagrammatic Classifications of Birds, 1819–1901: Views of the Natural System in Nineteenth-Century British Ornithology." In H. Quellet (ed.). *Acta xix Congressus Internationalis Ornithologici*, 2746–2759. Ottawa: National Museum of Natural Sciences.

———. 1991. "Representations of the Natural System in the Nineteenth Century." *Biology and Philosophy* 6: 255–274.

———. 1992. "Telling the Tree. Narrative Representation and the Study of Evolutionary History." *Biology and Philosophy* 7: 135–160.

Oldroyd, David R. 1984. "How Did Darwin Arrive at His Theory? The Secondary Literature to 1982." *History of Science* 22: 325–374.

Oppenheimer, Jane M. 1967. *Essays in the History of Embryology and Biology.* Cambridge, Mass.: MIT Press.

Ormond, Richard. 1981. *Sir Edwin Landseer.* New York: Rizzoli.

Ospovat, Dov. 1981. *The Development of Darwin's Theory: Natural History, Natural Theology, and Natural Selection, 1838–1859.* Cambridge, Mass.: Cambridge University Press.

Owen, Richard. 1848. *On the Archetype and Homologies of the Vertebrate Skeleton.* London: John van Voorst.

———. 1849a. *On the Nature of Limbs.* London: John van Voorst.

———. 1849b. "Osteological Contributions to the Natural History of the Chimpanzees (Troglodytes, Geoffroy), Including the Descriptions of the Skull of a Large Species (Troglodytes Gorilla, Savage) Discovered by Thomas S. Savage, MD in the Gaboon Country, West Africa." *Transactions of the Zoological Society of London* 3: 381–422.

———. 1860. *Paleontology: Or a Systematic Summary of Extinct Animals and Their Geological Relations.* Edinburgh: Adam and Charles Black.

Paley, William. 1802. *Natural Theology, Or Evidences of the Existence and Attributes of the Deity: Collected from the Appearances of Nature.* London: R. Faulder and Son.

———. 1836. *Natural Theology. With illustrative Notes by Henry Brougham, and Charles Bell.* London: C. Knight, 1936.

Palmer, Alfred H. 1895. *The Life of Joseph Wolf: Animal Painter.* London: Longmans.

Panofsky, Erwin. 1962. S*tudies in Iconology: Humanistic Themes in the Art of the Renaissance.* New York: Harper and Row.

Paradis, James G. 1997. "Satire and Science in Victorian Culture." In Bernhard Lightman (ed.). *Victorian Science in Context,* 143–175. Chicago: University of Chicago Press.

Patterson, David K. 1974. "Paul B. Du Chaillu and the Exploration of Gabon, 1855–1865." *International Journal of African Historical Studies* 4: 647–667.

Paul, Diane. 2003. "Darwin, Social Darwinism and Eugenics." In Jonathan Hodge and Gregory Radick (eds.). *The Cambridge Companion to Darwin,* 214–239. Cambridge: Cambridge University Press.

Peirce, Charles Sanders. 1893. "Evolutionary Love." *Monist* 3: 176–200.

Porter, Duncan M. 1982. "The *Beagle* Collector and His Collection." In David Kohn (ed.). *The Darwinian Heritage,* 973–1020. Prince-

ton, N.J.: Princeton University Press in association with Nova
Pacifica.

——. 1993. "On the Road to the *Origin* with Darwin, Hooker, and
Gray." *Journal of the History of Biology* 26: 1–38.

Prange, Wolfgang. 2002. "Der Hund des Goltzius. Das Kunstwerk
als Geschichtsdokument." *Weltkunst* 72: 38–40.

Prodger, Phillip. 1998a. *An Annotated Catalogue of the Illustrations
of Human and Animal Expression from the Collection of Charles
Darwin: An Early Case of the Use of Photography in Scientific
Research.* Lewiston, N.Y.: Edwin Mellen.

——. 1998b. "Illustration as Strategy in Charles Darwin's
'Expression of the Emotions in Man and Animals.'" In Timothy
Lenoir (ed.). *Inscribing Science: Materiality and Text in Scientific
Communication,* 140–182. Stanford, Calif.: Stanford University
Press.

——. 1998c. "Photography and the 'Expression of the Emo-
tions.'" In Charles Darwin. *The Expression of the Emotions in
Man and Animals.* Edited by Paul Ekman, 399–423. London:
Harper Collins.

——. 2009. *Darwin's Camera: Art and Photography in the Theory
of Evolution.* Oxford: Oxford University Press.

Quammen, David. 1996. T*he Song of the Dodo: Island Biogeography
in the Age of Extinction.* New York: Simon and Schuster.

Raby, Peter. 2001. *Alfred Russel Wallace. A Life.* London: Chatto
and Windus.

Rachootin, Stan P. 1985. "Owen and Darwin Reading a Fossil: Mac-
rauchenia in a Boney Light." In David Kohn (ed.). *The Darwinian Heritage,* 155–184. Princeton, N.J.: Princeton University
Press in association with Nova Pacifica.

Ratcliff, Marc J. 1999. "Temporality, Sequential Iconography and
Linearity in Figures: The Impact of the Discovery of Division
in Infusoria." *History and Philosophy of the Life Sciences* 21:
255–292.

Raulff, Ullrich, and Gary Smith (eds.). 1999. *Wissensbilder. Strategien der Überlieferung.* Berlin: Akademie Verlag.

Reade, William Winwood. 1863. *Savage Africa: Being the Narrative
of a Tour of Equatorial, South-western, and North-western Africa.*
London: Smith, Elder.

Rehbock, Philip F. 1983. *The Philosophical Naturalists: Themes in
Early Nineteenth-Century British Biology.* Madison: University of
Wisconsin Press.

Reichholf, Josef H. 2003. "Die kontingente Evolution. Entstehung und Entwicklung der Organismen." In Ernst Peter Fischer and Klaus Wiegandt (eds.). *Evolution. Geschichte und Zukunft des Lebens*, 45–75. Frankfurt am Main: Fischer Taschenbuch Verlag.

Rheinberger, Hans-Jörg. 1986. "Aspekte des Bedeutungswandels im Begriff organismischer Ähnlichkeit vom 18. zum 19. Jahrhundert." *History and Philosophy of the Life Sciences* 8: 237–250.

———. 1997. *Toward a History of Epistemic Things: Synthesizing Proteins in the Test Tube*. Stanford, Calif.: Stanford University Press.

———. 2003a. "Historische Beispiele experimenteller Kreativität in den Wissenschaften." In Walter Berka, Emil Brix, and Christian Smekal (eds.). *Woher kommt das Neue? Kreativität in Wissenschaft und Kunst*, 29–49. Vienna: Böhlau.

———. 2003b. "Die Politik der Evolution. Darwins Gedanken in der Geschichte." In Ernst Peter Fischer and Klaus Wiegandt (eds.). *Evolution. Geschichte und Zukunft des Lebens*, 178–197. Frankfurt am Main: Fischer Taschenbuch Verlag.

———. 2003c. "Präparate—'Bilder' ihrer selbst. Eine bildtheoretische Glosse." *Bildwelten des Wissens* 2: 9–19.

———, and Peter McLaughlin. 1984. "Darwin's Experimental Natural History." *Journal of the History of Biology* 17: 345–368.

———, Michael Hagner, and Bettina Wahrig-Schmidt (eds.). 1997. *Räume des Wissens. Repräsentation, Codierung, Spur*. Berlin: Akademie Verlag.

Rhodes, Frank H. T. 1991. "Darwin's Search for a Theory of the Earth: Symmetry, Simplicity, and Speculation." *British Journal for the History of Science* 24: 193–229.

Rhodes, Richard. 2004. *John James Audubon. The Making of an American*. New York: Alfred A. Knopf.

Rice, Tony. 1999. *Voyages of Discovery: Three Centuries of Natural History Exploration*. New York: Potter.

Richards, Eveleen. 1976. "The German Concept of Embryonic Repetition and Its Role in Evolutionary Theory in England up to 1859." PhD diss., University of New South Wales.

———. 1987. "A Question of Property Rights: Richard Owen's Evolutionism Reassessed." *British Journal for the History of Science* 20: 129–171.

Richards, Robert J. 1992. *The Meaning of Evolution. The Morphological Construction and Ideological Reconstruction of Darwin's Theory*. Chicago: University of Chicago Press.

———. 2008. *The Tragic Sense of Life. Ernst Haeckel and the Struggle over Evolutionary Thought.* Chicago: Chicago University Press.

Ridgway, Robert. 1889. "Scientific Results of Explorations by the U.S. Fish Commission Steamer Albatross. No. 1.—Birds Collected on the Galapagos Islands in 1888." *Proceedings of the U.S. National Museum* 12: 101–139.

Ritvo, Harriet. 1987. *The Animal Estate: The English and Other Creatures in the Victorian Age.* Cambridge, Mass.: Harvard University Press.

———. 1997. *The Platypus and the Mermaid. And Other Figments of the Classifying Imagination.* Cambridge, Mass.: Harvard University Press.

Roberts, Michael. 1997. "Darwin's Doubts About Design. The Darwin-Gray-Correspondence." *Science and Christian Belief* 9: 113–127.

Root, Nina J., and R. Bryan Johnson. 1986a. *Transactions of the Zoological Society of London. An Index to the Artists 1835–1936.* New York: Garland.

———. 1986b. *Proceedings of the Zoological Society of London. An Index to the Artists 1848–1900.* New York: Garland.

Rubin, James H. 2003. *Impressionist Cats and Dogs. Pets in the Painting of Modern Life.* New Haven: Yale University Press.

Rudwick, Martin J. S. 1975. "Caricature as a Source for the History of Science: De la Beche's Anti-Lyellian Sketches of 1831." *Isis* 66: 534–60.

———. 1976. "The Emergence of a Visual Language for Geological Science, 1760–1840." *History of Science* 14: 149–195.

———. 1982. "Charles Darwin in London. The Integration of Public and Private Science." *Isis* 73: 186–206.

———. 1992. *Scenes from Deep Time: Early Pictorial Representations of the Prehistoric World.* Chicago: University of Chicago Press.

———. 2000. "Georges Cuvier's Paper Museum of Fossil Bones." *Archives of Natural History* 27: 51–68.

Rupke, Nicolaas A. (ed.). 1987. *Vivisection in Historical Perspective.* London: Croom Helm.

———. 1994. *Richard Owen: Victorian Naturalist.* New Haven: Yale University Press.

———. 1995. "Richard Owen: Evolution ohne Darwin." In Eve-Marie Engels (ed.). *Die Rezeption der Evolutionstheorien im 19. Jahrhundert,* 214–224. Frankfurt am Main: Suhrkamp.

Ruse, Michael. 1979. *The Darwinian Revolution: Science Red in Tooth and Claw.* Chicago: University of Chicago Press.

———. 2003. *Darwin and Design: Does Evolution Have a Purpose?* Cambridge, Mass.: Harvard University Press.

Ruskin, John. 1906. "Lectures on Landscape: Ten Lectures on the Relation of the Natural Sciences to Art." In E. T. Cook and Alexander Wedderburn (eds.). *The Works of John Ruskin.* Vol. 22. London: Allen.

Russell, Edward Stuart. 1916. *Form and Function: A Contribution to the History of Animal Morphology.* London: Murray.

Salvin, Osbert. 1876. "On the Avifauna of the Galapagos Archipelago." *Transactions of the Zoological Society* 9: 447–510.

Sauer, Gordon C. 1982. *John Gould, the Bird Man: A Chronology and Bibliography.* Melbourne, N.Y.: Lansdowne Editions.

Savage, Thomas S., and Jeffries Wyman. 1847. "A Description of the External Characters and Habits of Troglodytes Gorilla, a New Species of Orang from the Gaboon River, and of the Osteology of the Same." *Boston Journal of Natural History* 5: 417–443.

Schlögl, Uwe. 1999. "Vom Frosch zum Dichter-Apoll." In Gerda Mraz and Uwe Schlögl (eds.). *Das Kunstkabinett des Johann Caspar Lavater* (exhibition catalog), 164–171. Vienna: Böhlau.

Schmidt, Dietmar. 2003. "'Viehsionomik.' Repräsentationsformen des Animalischen im 19. Jahrhundert." *Historische Anthropologie* 11: 21–46.

Schmidt-Burkhardt, Astrit. 2005. *Stammbäume der Kunst. Zur Genealogie der Avantgarde.* Berlin: Akademie Verlag.

Schmölders, Claudia. 1995. *Das Vorurteil im Leibe. Eine Einführung in die Physiognomik.* Berlin: Akademie Verlag.

Schulze-Hagen, Karl, and Armin Geus (eds.). 2000. *Joseph Wolf (1820–1899): Tiermaler/ Joseph Wolf (1820–1899): Animal Painter.* Parallel German and English texts. Marburg an der Lahn: Basilisken.

———, Frank Steinheimer, Ragnar Kinzelbach, and Christoph Gasser. 2003. "Avian Taxidermy in Europe from the Middle Ages to the Renaissance." *Journal of Ornithology* 144: 459–478.

Secord, James A. 1981. "Nature's Fancy. Charles Darwin and the Breeding of Pigeons." *Isis* 72: 163–186.

———. 1985. "Darwin and the Breeders. A Social History." In David Kohn (ed.). *The Darwinian Heritage,* 519–542. Princeton, N.J.: Princeton University Press in association with Nova Pacifica.

————. 1991a. "The Discovery of a Vocation. Darwin's Early Geology." *British Journal for the History of Science* 24: 133–157.

————. 1991b. "Edinburgh Lamarckians: Robert Jameson and Robert E. Grant." *Journal of the History of Biology* 24: 1–18.

————. 2000. *Victorian Sensation: The Extraordinary Publication, Reception, and Secret Authorship of "Vestiges of the Natural History of Creation."* Chicago: University of Chicago Press.

Secord, William. 2002. *Dog Painting: The European Breeds.* Woodbridge: Antique Collector's Club.

Serpell, James. 2005. "People in Disguise: Anthropomorphism and the Human-Pet Relationship." In Lorraine Daston and Gregg Mitman (eds.). *Thinking with Animals. New Perspectives on Anthropomorphism,* 121–136. New York: Columbia University Press.

Shapin, Steven. 1989. "The Invisible Technician." *American Scientist* 77: 545–563.

Shermer, Michael. 2002. *In Darwin's Shadow: The Life and Science of Alfred Russel Wallace. A Biographical Study on the Psychology of History.* Oxford: Oxford University Press.

Sloan, Phillip R. 1985. "Darwin's Invertebrate Program, 1826–1836: Preconditions for Transformism." In David Kohn (ed.). *The Darwinian Heritage,* 71–120. Princeton, N.J.: Princeton University Press in association with Nova Pacifica.

Smith, Christopher U. M., and Robert Arnott. 2004. *The Genius of Erasmus Darwin.* Aldershot: Ashgate.

Smith, Jonathan. 2001. "John Gould, Charles Darwin, and the Picturing of Natural Selection." *The Book Collector* 50: 51–76.

————. 2006. *Charles Darwin and Victorian Visual Culture.* Cambridge: Cambridge University Press.

Smith, Lindsay. 1995. *Victorian Photography, Painting and Poetry: The Enigma of Visibility in Ruskin, Morris and the Pre-Raphaelites.* Cambridge: Cambridge University Press.

Soemmerring, Samuel Thomas von. 1799. *Icones embryonum humanorum.* Frankfurt am Main: Varrentrapp and Wenner.

Spary, Emma. 1999. "Codes of Passion. Natural History Specimens as a Polite Language in Late Eighteenth-Century France." In Hans Erich Bödeker, Hanns Reill, and Jürgen Schlumbohm (eds.). *Wissenschaft als kulturelle Praxis. 1750–1900,* 105–135. Göttingen: Vandenhoeck and Ruprecht.

Spencer, Frank. 1995. "Pithekos to Pithecanthropus: An Abbreviated Review of Changing Scientific Views on the Relationship of the Anthropoid Apes to Homo." In Raymond Corbey and Bert

Theunissen (eds.). *Ape, Man, Apeman. Changing Views Since 1600*. Evaluative proceedings of the symposium. Leiden, the Netherlands, 28 June–1 July 1993, 13–28. Leiden: Leiden University.

Spencer, Herbert. 1860. "The Physiology of Laughter." *Macmillan's Magazine* 1: 395–402.

Stafford, Barbara Maria. 1984. *Voyage into Substance. Art, Science, Nature, and the Illustrated Travel Account, 1760–1840*. Cambridge, Mass.: MIT Press.

——. 1994. *Artful Science: Enlightenment Entertainment and the Eclipse of Visual Education*. Cambridge, Mass.: MIT Press.

Star, Susan Leigh, and James R. Griesemer. 1989. "Institutional Ecology, 'Translations' and Boundary Objects: Amateurs and Professionals in Berkeley's Museum of Vertebrate Zoology, 1907–1939." *Social Studies of Science* 19: 387–420.

Stauffer, Robert C. (ed.). 1975. *Charles Darwin's "Natural Selection," Being the Second Part of His Big Species Book Written from 1856 to 1858*. London: Cambridge University Press.

Steinheimer, Frank. 2004. "Charles Darwin's Bird Collection and Ornithological Knowledge During the Voyage of H.M.S. *Beagle*, 1831–1836." *Journal of Ornithology* 145: 300–320.

——, E. C. Dickinson, and M. Walters. 2006. "The Zoology of the Voyage of H.M.S. *Beagle*, Part III, Birds: New Avian Names, Their Authorship and Their Dates." *Bulletin of the British Ornithologists' Club* 126: 171–193.

——, and W. Sudhaus, 2006. "Die Speziation der Darwinfinken und der Mythos ihrer initialen Wirkung auf Charles Darwin." *Naturwissenschaftliche Rundschau* 59: 409–422.

Stepan, Nancy Leys. 2001. *Picturing Tropical Nature*. London: Reaktion.

Sternberger, Dolf. 1977. *Panorama of the Nineteenth Century*. New York: Urizen.

Storch, Volker, Ulrich Welsch, and Michael Wink. 2001. *Evolutionsbiologie*. Berlin: Springer.

Stott, Rebecca. 2003. *Darwin and the Barnacle. The Story of One Tiny Creature and History's Most Spectacular Scientific Breakthrough*. London: Faber.

Stresemann, Erwin. 1951. *Ornithology from Aristotle to the Present*. Cambridge, Mass.: Harvard University Press.

Strickberger, Monroe W. 2000. *Evolution*, 3. Sudbury, Mass.: Jones and Bartlett.

Strickland, Hugh Edwin. 1835. "On the Arbitrary Alteration of Established Terms in Natural History." *Magazine of Natural History* 8: 36–40.

———. 1841. "On the True Method of Discovering the Natural System." *Annals and Magazine of Natural History* 6: 184–194.

Sulloway, Frank. 1982a. "Darwin and His Finches. The Evolution of a Legend." *Journal of the History of Biology* 15: 1–53.

———. 1982b. "Darwin's Conversion: the *Beagle* Voyage and Its Aftermath." *Journal of the History of Biology* 15: 325–96.

———. 1982c. "The *Beagle* Collections of Darwin's Finches (Geospizinae)." *Bulletin of the British Museum (Natural History), Zoology Series* 43: 49–94.

Swainson, William. 1822. *The Naturalist's Guide for Collecting and Preserving Subjects of Natural History and Botany, Both in Temperate and Tropical Countries.* London: W. Wood.

———. 1835. *A Treatise on the Geography and Classification of Animals.* London: Longman.

———. 1836–1837. *The Natural History and Classification of Birds.* 2 vols. London: Longman.

———. 1840. *Taxidermy: With the Biography of Zoologists, and Notices of Their Works.* London: Longman.

Swoboda, Gudrun. 1999. "Die Sammlung Johann Caspar Lavater in Wien. Herkunft, Struktur, Funktion." In Gerda Mraz and Uwe Schlögl (eds.). *Das Kunstkabinett des Johann Caspar Lavater* (exhibition catalog), 74–95. Vienna: Böhlau.

Tegetmeier, William. 1871. "The Argus Pheasant." *The Field,* January 14, 20.

———. 1874. "Display of the Argus Pheasant." *The Field,* March 28, 296.

Tree, Isabella. 1991. *The Ruling Passion of John Gould. A Biography of the Bird Man.* London: Barrie and Jenkins.

Uglow, Jenny. 2006. *Nature's Engraver: A Life of Thomas Bewick.* London: Faber.

Ullrich, Wolfgang. 2002. *Die Geschichte der Unschärfe.* Berlin: Wagenbach.

Uschmann, Georg. 1967. "Zur Geschichte der Stammbaum-Darstellungen." In Manfred Gersch (ed.). *Gesammelte Vorträge über moderne Probleme der Abstammungslehre.* Vol. 2, 9–30. Jena: Friedrich-Schiller-Universität.

Van Wyhe, John. 2007. "Mind the Gap: Did Darwin Avoid Publishing His Theories for Many Years?" *Notes and Records of the Royal Society* 61: 177–205.

Vaucaire, Michel. 1930. *Paul Du Chaillu, Gorilla Hunter: Being the Extraordinary Life and Adventures of Paul Du Chaillu*. New York: Harper.

Vigors, Nicholas Aylward. 1838. "On the Natural Affinities That Connect the Orders and Families of Birds." *Transactions of the Linnean Society* 24: 395–517.

Voland, Eckart, and Karl Grammer (eds.). 2003. *Evolutionary Aesthetics*. Berlin: Springer.

Voss, Julia. 2003a. *Darwins Diagramme. Bilder von der Entdeckung der Unordnung*. Preprint des Max-Planck-Instituts für Wissenschaftsgeschichte, No. 249, Berlin.

———. 2003b. "Augenflecken und Argusaugen. Zur Bildlichkeit der Evolutionstheorie." *Bildwelten des Wissens* 1: 75–85.

———. 2004. "Das erste Bild der Evolution. Wie Charles Darwin die Unordnung der Naturgeschichte zeichnete und was daraus wurde." In Bernhard Kleeberg, Wolfgang Lefèvre, and Julia Voss (eds.). *Zum Darwinismusstreit*. Preprint des Max-Planck-Instituts für Wissenschaftsgeschichte, No. 272, Berlin.

———. 2008. "Darwin oder Moses? Funktion und Bedeutung von Charles Darwins Porträt im 19. Jahrhundert." *NTM Zeitschrift für Geschichte der Wissenschaften, Technik und Medizin* 16: 213–243.

———. 2009a. "Variation and Selection: The Theory of Evolution in the English and German Illustrated Press of the Nineteenth Century." In Pamela Kort and Max Hollein (eds.). *Darwin and the Search for Origins*, 246–257. Frankfurt a. M.: Wienand Verlag.

———. 2009b. "Monkeys, Apes and Evolutionary Theory: From Human Descent to King Kong." In Diana Donald and Jane Munro (eds.). *Endless Forms. Charles Darwin, Natural Science and the Visual Arts*, 215–234. New Haven: Yale University Press.

Wallace, Alfred Russel. 1855. "On the Law Which Has Regulated the Introduction of New Species." In *Contributions to the Theory of Natural Selection. A Series of Essays*, 1–25. London: Macmillan.

———. 1856. "Attempts at a Natural Arrangement of Birds." *Annals and Magazine of Natural History, 2nd ser.* 18: 193–216.

———. 1858. "On the Tendency of Varieties to Depart Indefinitely from the Original Type." In *Contributions to the Theory of Natural Selection. A Series of Essays*, 26–44. London: Macmillan.

———. 1869. *The Malay Archipelago: The Land of the Orang-Utan and the Bird of Paradise. A Narrative of Travel, with Studies of Man and Nature*. London: Macmillan.

Wallis, Brian. 1995. "Black Bodies, White Science: Louis Agassiz' Slave Daguerreotypes." *American Art* 9: 39–61.

Weigel, Sigrid. 2003. "Genealogie. Zur Ikonographie und Rhetorik einer epistomologischen Figur in der Geschichte von Kultur- und Naturwissenschaft." In Helmar Schramm et al. (eds.). *Bühnen des Wissens. Interferenzen zwischen Wissenschaft und Kunst,* 226–267. Berlin: Dahlem University Press.

Weiner, Jonathan. 1994. *The Beak of the Finch: A Story of Evolution in Our Time.* New York: Knopf.

Weinshank, Donald J., S. J. Ozminski, P. Ruhlen, and W. M. Barrett. 1990. *A Concordance to Charles Darwin's Notebooks, 1836–1844.* Ithaca, N.Y.: Cornell University Press.

Wellmann, Janina. 2003. "Wie das Formlose Formen schafft: Bilder in der Haller-Wolff-Debatte und die Anfänge der Embryologie um 1800." *Bildwelten des Wissens* 1: 105–115.

Whewell, William. 1837. *History of the Inductive Sciences, from the Earliest to the Present Time.* Vol. 3. London: J. W. Parker.

———. 2000 [1846]. *Indications of the Creator: Extracts, Bearing upon Theology, from the History and the Philosophy of the Inductive Sciences.* Reprint of the second edition, J. M. Lynch (ed.). Bristol: Thoemmes.

White, Paul. 2003. *Thomas Huxley: Making the "Man of Science."* Cambridge: Cambridge University Press.

———. 2005. "The Experimental Animal in Victorian Britain." In Lorraine Daston and Gregg Mitman (eds.). *Thinking with Animals. New Perspectives on Anthropomorphism,* 59–81. New York: Columbia University Press.

———. 2006. "Sympathy Under the Knife: Experimentation and Emotion in late Victorian Medicine." In Fay Bound Alberti (ed.). *Medicine, Emotion and Disease, 1700–1950,* 100–124. London: Palgrave Macmillan.

Winsor, Mary P. 1976. *Starfish, Jellyfish and the Order of Life: Issues in Nineteenth-Century Science.* New Haven: Yale University Press.

———. 1991. *Reading the Shape of Nature. Comparative Zoology at the Agassiz Museum.* Chicago: University of Chicago Press.

Wollschläger, Hans. 2002. *Tiere sehen dich an oder das Potential Mengele.* Göttingen: Wallstein.

Wood, John George. 1862. *The Illustrated Natural History.* London: Routledge.

———. 1869. *Bible Animals. Being a Description of Every Living Creature Mentioned in the Scriptures from the Ape to the Coral.* London: Routledge.

Wood, Marcus. 2000. *Blind Memory. Visual Representations of Slavery in England and America 1780–1856.* New York: Routledge.

Wood, Theodore. 1890. *The Rev. J. G. Wood: His Life and Work.* London: Cassell.

Woolner, Amy. 1917. *Thomas Woolner: Sculptor and Poet. His Life in Letters.* London: Chapman and Hall.

Wullen, Moritz (ed.). 2004. *Natur als Vision. Meisterwerke der Englischen Präraffaeliten* (exhibition catalog). Berlin: SMB-Dumont.

Yanni, Carla. 1999. *Nature's Museums: Victorian Science and the Architecture of Display.* London: Athlone.

Young, David. 1992. *The Discovery of Evolution.* Cambridge: Natural History Museum in association with Cambridge University Press.

Young, Robert C. 1995. *Colonial Desire. Hybridity, Culture and Race.* London: Routledge.

Zapperi, Roberto. 2004. *Der wilde Mann von Teneriffa: Die wundersame Geschichte des Pedro Gonzalez und seiner Kinder.* Munich: Beck.

Zimmerman, Hans Joachim. 1991. *Der akademische Affe. Die Geschichte einer Allegorie aus Cesare Ripas "Iconologia."* Wiesbaden: Reichert.

INDEX

actualism, 46
Adam and Eve (Dürer), 185, 186
adaptive radiation, 60
aestheticism: in nature, 256–258,
  297n17, 298n20
aesthetic preferences: of the
  argus pheasant, 135, 172–173,
  174–175
Agassiz, Louis, 69, 104, 105–108,
  120, 141, 210–211, 224, 254,
  278–279n66, 279nn68–69
Albert, Prince, 153, 200
Ali, Mehmet, 23
All the Year Round, 202
animals: attributes of associated
  with human physiognomy,
  190–192, 289–290n17; emotions
  as manifested in, 188–195,
  213–226; humans' kinship to,
  143–145, 187–189, 217, 246–248;
  illustrations of, 213–226;
  laughter of, 244–247; portraits
  of, 218–219, 293n65; protection
  of, 219. *See also* apes; birds;
  gorillas; pets
animal shelters, 219
Anthropogenie (Haeckel), 122–
  123, 145, 178, 179–180, 182–183

antivivisection movement, 219,
  293n66
ant lions, 46–47
apes: associated with evolutionary theory, 227–228; Darwin's
  view of, 181–182, 243–244;
  emotions manifested by,
  244–247; kinship of with
  humans, 143–145, 147, 182–187,
  226–228, 234–238, 243,
  283–284n27. *See also* gorillas
archetype theory, 3, 136–138,
  142–143, 164
argus pheasant, 6, 7; aesthetic
  preferences of, 135, 172–173,
  174–175; as argument against
  evolution, 162–167; as evidence
  of chance variation, 149–150,
  167–173, 176–177, 254; illustrations of, 127–135, 166–173;
  mating dance of, 135, 172–174;
  as subject of evolutionary
  research, 133–135, 149–150,
  166–174, 175–180; trompe l'oeil
  effect on wings of, 131, 164,
  169, 175
Argyll, George Douglas Campbell, Duke of, 143; Darwin's